629.287
I19
1999

IDENTIFICATION

PARTS FAILU

A highly-illustrated failure analysis guide for automotive an

629.287 I19 1999
Identification of parts
failures : a
highly-illustrated failure
analysis guide for
automotive and off-road
vehicle parts
50852

050852

FUNDAMENTALS OF SERVICE

PUBLISHER: DEERE & COMPANY
JOHN DEERE PUBLISHING
Almon TIAC Building, Suite 104
1300 – 19th Street
East Moline, IL 61244

http://www.deere.com/aboutus/pub/jdpub
1-800-522-7448

JOHN DEERE PUBLISHING EDITORIAL STAFF:
Lori J. Dhabalt
Cindy S. Calloway

TO THE READER

This manual shows many pictures of failed parts. A brief explanation accompanies each picture and gives reasons why the part might have failed. These causes are not necessarily the only conditions that could cause the part to fail. All factors concerning the operation and maintenance of a machine must be considered when analyzing failures. Recommended actions are given for each failed part.

In most cases the failures shown are secondary failures — some other problem may have caused the failure of the part that is pictured.

It is important to find the original problem area when diagnosing failures of any part.

This manual cannot identify all causes of failed parts. Where the cause of the failure cannot be determined in the field, the part may require examination by a qualified materials engineer. Information on service and operation should be provided with the failed part.

Copyright © 1978, 1979, 1987, 1991, 1999. Deere & Company, John Deere Publishing. All Rights Reserved

This material is the property of Deere & Company, John Deere Publishing, all use and/or reproduction not specifically authorized by Deere & Company, John Deere Publishing is prohibited.

We have a
long-range interest in
good service

ISBN 0-86691-266-5

ACKNOWLEDGEMENTS

John Deere gratefully acknowledges the writing and editing expertise of Robert Gunter and William Rockstroh. We also appreciate the efforts of the following groups: Professor Charles Lipson, University of Michigan; American Gear Manufacturers Association; American Society for Metals; Cummins Engine Company, Inc.; Dana Corporation; New Departure - Hyatt Bearings, Division of General Motors; Eaton Corporation; Garrett Automotive Group, Division of Allied Signal, Inc.; The Gates Rubber Company; Case - International; Lipson & Colwell, *HANDBOOK OF MECHANICAL WEAR*, 1961. University of Michigan Press; *MACHINE DESIGN MAGAZINE*, October 13, 1966, article by Henry Suss, General Electric Co.; Rex Chainbelt, Inc.; SKF Industries, Inc.; E.E Shipley; Society of Automotive Engineers, Inc.; Society for Experimental Stress Analysis; Sundstrand Hydrotransmission; TRW, Inc.; Timken Roller Bearing Co.; Donald J. Wulpi, *HOW COMPONENTS FAIL*, 1966, American Society for Metals. **Thanks also to a host of John Deere people for their valuable suggestions and comments.**

FUNDAMENTALS OF SERVICE (FOS) is a series of manuals created by Deere & Company. Each book in the series is conceived, researched, outlined, edited, and published by Deere & Company, John Deere Publishing. Authors are selected to provide a basic technical manuscript which is edited and rewritten by staff editors.

FOR MORE INFORMATION

This book is one of many books published on agricultural and industrial machinery. For more information or to request a FREE CATALOG, call 1-800-522-7448 or send your request to address above or

Visit Us on the Internet--
http://www.deere.com/aboutus/pub/jdpub/

CONTENTS

1

PISTONS, RINGS, CYLINDER LINERS AND GASKETS

2 JOURNAL BEARINGS

3 VALVE GEAR TRAIN

4 TURBOCHARGERS

5 GEARS

6

SHAFTS, AXLES, SPINDLES, AND UNIVERSAL JOINTS

7

HYDROSTATIC TRANSMISSIONS

8

ANTI-FRICTION BEARINGS

9

BELTS AND CHAINS

10 TRACKS AND TIRES

11 MISCELLANEOUS FAILURES

APPENDIX

PISTONS, RINGS, CYLINDER LINERS AND GASKETS

1

INTRODUCTION

Pistons, piston rings and cylinder liners are an essential part of an engine:

- **Pistons receive the force of combustion and transmit it through the connecting rods to the crankshaft.**
- **Cylinder liners guide the piston.**
- **Piston rings form a gas-tight seal between the piston and the cylinder.**

This section deals with identifying failures of these following components:

- **Pistons — gasoline engines**
- **Pistons — diesel engines**
- **Piston rings**
- **Cylinder liners**
- **Gaskets**

IDENTIFYING PISTON FAILURES IN GASOLINE ENGINES

Although some failures in pistons operating in gasoline engines are the same as in diesel engines, there are enough differences to justify separate groupings.

The principal causes of piston failures in gasoline engines are:

- **Detonation**
- **Preignition**
- **Scuffing and scoring**
- **Corrosive wear**
- **Physical damage to pistons**

DETONATION

Detonation is uncontrolled combustion accompanied by a loss of power and waste of energy. The piston is frequently damaged.

These pistons have been damaged by detonation. Because of the hammering pressures, piston damage usually appears as fractures on or through the crown, or in the skirt and piston pin area (large arrows).

The top piston land can be crushed causing the ring to bind in the ring groove.

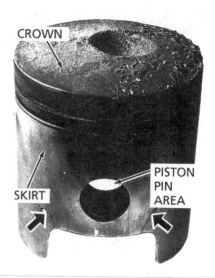

Fig. 1

Fig. 2

A noticeable knocking occurs when fuel in the cylinder ignites too early, too rapidly, or unevenly. The resulting "knock" can burn the piston, wear out the top groove, or cause the ring to break or stick.

Causes of combustion knock:

- **Fuel octane too low**
- **Lean fuel mixtures**
- **Ignition timing advanced too much**
- **Lugging the engine or overfueling**
- **Cooling system not working (overheating)**

Recommendation: Replace

Fig. 3

Fig. 4

PREIGNITION

Occurs when the fuel ignites before the spark occurs. As a result, part of the fuel burns while the piston is still coming up on its compression stroke. The burning fuel is compressed and overheated by the piston, and by further combustion. The heat can get so intense that engine parts melt.

These pistons were damaged by the heat of preignition. The intense heat of preignition burned and melted the piston. Damage may appear on and through the crown, through the ring lands, or both.

Fig. 5

CAUSES OF PREIGNITION:

- **Carbon deposits that remain hot enough to ignite fuel early** .

- **Overheating**

- **Valve operating too hot because of excessive guide clearance or bad seats**

- **Hot spots caused by damaged rings**

- **Spark plugs (improper heat range or reach)**

- **A loose spark plug**

Recommendation: Replace

Fig. 6

SCUFFING AND SCORING

Scuffing and scoring (adhesive wear) is caused by too much heat. When two metal parts rub and the heat builds up to the welding point, a small deposit or "hot spot" of metal is pulled out and deposited on the cooler surface.

Scuffing leaves discolored areas on the surface of rings, pistons and cylinder walls.

Scuffing starts as tiny surface disturbances. If they are not removed, scuffing spreads and becomes noticeable and more severe. It is then called scoring. Any engine condition which heats rubbing parts to the welding point, or which prevents the transfer of heat from these surfaces, influences scuffing.

The following are possible causes of scuffing and scoring:

- **Improper warm-up**

- **Lubricating system not functioning**

- **Cooling system plugged**

- **Combustion knock and preignition**

- **Lugging or overloading**

- **Misaligned connecting rod**

Fig. 7

Occasionally an unusual wear pattern results from a mis-aligned connecting rod. Contact with cylinder wall shows on the bottom of the skirt at left, and at the ring lands on the right (arrows). There is also a diagonal wear pattern extending across the skirt of the piston. This uneven wear is due to a bent or twisted connecting rod or shaft. When connecting rods are misaligned, rings do not have proper contact with cylinder wall, pistons wear rapidly and unevenly, oil consumption is excessive and the engine is prone to scuff or score. Always check rod alignment.

Recommendation: Replace

Fig. 8

CORROSIVE WEAR

Corrosive wear shows up as spotted grayish pitted surface on pistons or cylinder walls.

The following are possible causes of corrosive wear:

- **Leaking coolant**
- **Cold engine operation or putting engine under load before it has reached operating temperature**
- **Wrong lubricating oil or dirty oil**
- **Acids resulting from combustion or the reaction of moisture and sulfur in the lubricating oil**

Other corrosion may be harder to find. If excessive wear is found, and scuffing and scoring are eliminated as causes, suspect corrosive wear.

Recommendation: Replace

Fig. 9

PHYSICAL DAMAGE TO PISTONS

Physical damage to pistons can be caused by:

- **The piston losing its pin lock (illustrated)**
- **Connecting rod out of alignment**
- **Crankshaft with too much endplay**
- **Crankshaft journal with too much taper**
- **Cylinder bore out of alignment**
- **Improperly installed piston pin locks**
- **Ring groove scratched while cleaning out carbon**
- **Piston handled carelessly or dropped**

Recommendation: Replace

Fig. 10

IDENTIFYING PISTON FAILURES IN DIESEL ENGINES

The principal causes of piston failures in diesel engines are:

- **Cracking**
- **Breakage**
- **Wear**
- **Scuffing and seizure**
- **Erosion**

CRACKING

Small hairline cracks will appear in the piston crater after normal operation. A piston with hairline cracks should be replaced.

When crater cracks are open, deep, or interconnected, the piston top has overheated.

Recommendation: Replace

Fig. 11 — Hairline Cracks

Fig. 12 — Severe Cracking

PISTON BREAKAGE

This piston was damaged by a broken glow plug tip. Tips break due to incorrect timing, shorting out, or rough combustion.

Recommendation: Replace

Fig. 13

A broken valve damaged this piston.

Recommendation: Replace

Fig. 14

A loose heat plug eventually broke leaving a battered hole in the crown where it was located. Note the impressions left by its threads in the top of the piston. Impressions give a clue to the cause of piston damage.

Fig. 15

On rare occasions the bond between the piston and the cast iron insert, which holds the top ring, will fail. This type of failure probably occurs because of intense piston crown heat due to a constant overload operation or extreme load fluctuation.

Recommendation: Replace

Fig. 16

The iron insert has loosened and broken the ring land below it.

Recommendation: Replace

Fig. 17

Lack of lubrication damaged this piston and connecting rod when the oil passage to the connecting rod became blocked. The blockage of the passage occurred when the main bearing of the crankshaft spun and seized up.

Recommendation: Replace

Fig. 18

Pistons, Rings, Cylinder Liners and Gaskets

Under constant pounding the loose insert caused the outer edge of the piston crown to break away.

Fig. 19

The broken insert continued to pound the piston, rounding off the edges of the insert retaining groove.

Recommendation: Replace

Fig. 20

Pistons usually break at the pin bore because of piston seizure — occasionally caused by overspeeding the engine. If piston seizure caused the breakage, the piston will bear extensive scuff marks (arrow). If the broken piston is virtually unmarked, then breakage at the pin could be attributed to overspeeding. Overspeeding almost always shows evidence of piston hitting valve. Ether overdose, advanced timing, and overfueling are most common causes of piston breakage.

Recommendation: Replace

Fig. 21

A flaked off piece of aluminum from the bonded steel piston insert ultimately caused the hole in this piston crown. The aluminum flake lodged between the piston and valve head stressing the valve stem which eventually broke causing the valve head to be embedded in the piston.

Recommendation: Replace

Fig. 22

WEAR

Wear produces a pattern similar to bearing wear — a shiny surface with many tiny scratches caused by the piston skirt occasionally contacting the liner. This slightly roughened surface aids in lubrication.

Recommendation: Reuse

However, if pistons have any of the following skirt damage mentioned, they should not be used again.

Fig. 23

Abrasive wear is indicated by a piston skirt of a very dull gray, oil ring rails worn away, chrome facings worn off all rings, grooves badly worn, and some liner wear. This shows a piston badly worn by abrasives. Pistons like this cannot be reused. Dirt probably entered through the air inlet system and mixed with lubricating oil. Check piston-ring and ring-groove dimensions against technical manual specifications whenever the above conditions appear on the piston. An engine which shows piston and ring wear like this should be checked for leaks in the air inlet system.

Recommendation: Replace

Fig. 24

SCUFFING AND SEIZURE

Scuffing and seizure are two related forms of piston damage: the first is usually mild and the second severely damaging.

Scuffing on the top ring land can result when the piston top overheats and expands beyond its normal size. With metal-to-metal contact, the softer aluminum of the piston will pull away and stick to the liner wall.

A leaking fuel valve or improper timing is the most common cause of the scuffing when it is confined to the first land. In an extreme case piston seizure will result.

Recommendation: Replace

Fig. 25

Scuffing in streaks on the piston skirt—particularly in the pin bore area — with little or no scuffing on the first land may be caused by inadequate engine cooling. If this type of damage appears on most or all of the pistons, cooling system failure or inadequate lubrication is the likely cause.

Check for:

- **Water or antifreeze leaks**

- **Crusty deposits in the ring lands, ring belt and crown of the piston**

- **Crankcase oil level**

- **Condition of the underside of the piston and the cooling jet**

- **Broken fan or water pump belt**

- **Plugged radiator**

Recommendation: Replace

Fig. 26

Small spots of skirt scuffing could be caused by abrasives from the crankcase oil, a bit of broken ring, or other material. Cold starts can also cause this type of scuffing, and it can occur due to inadequate lubrication in cold weather.

Recommendation: Replace

Fig. 27

Scuffing marks extending from the top to the bottom of the piston can indicate the piston was oversized. Cooling or lubrication system failures can also produce full length scuffing.

Recommendation: Replace

Fig. 28

EROSION

Erosion of the piston top is generally caused by rough combustion. When valve pockets are eroded, a leaking fuel valve or incorrect timing is the likely cause. When the fuel burning cycle is not normal, incomplete or "rough" combustion occurs. This causes high peak pressures and excessively high temperatures which could damage the piston crown. Check the fuel injection nozzles and engine timing.

Excessive heat can also be caused by a malfunctioning governor overspeeding the engine.

Recommendation: Replace

Fig. 29

Inadequate air supply can cause rough combustion resulting in erosion of the piston top.

Recommendation: Replace

Fig. 30

In this case, erosion is more uniform around the outer edge of the crater and is located away from the inlet valve pocket. Check for restrictions or leaks in the inlet air system.

Recommendation: Replace

Fig. 31

Here are other examples of crown erosion. The condition of the blackened piston on the top and the eroded piston on the bottom was caused by burning liquid fuel.

Fig. 32

Advanced timing or high altitude operation without derating the engine can cause fuel to collect on the top of the piston resulting in severe erosion. This leads to piston seizure.

Recommendation: Replace

Fig. 33

IDENTIFYING PISTON RING FAILURES

Piston rings do three jobs:

- **Form a gas-tight seal between the piston and cylinder**

- **Help cool the piston by transferring heat**

- **Control lubrication between piston and cylinder wall**

The principal causes for failure of piston rings are:

- **Wear**

- **Chipping**

- **Scuffing and scoring**

- **Breakage**

- **Sticking**

WEAR

When the ring faces are covered with dull gray, vertical scratches and have excessive ring-to-groove clearance, the rings have been worn by abrasives.

Other indications of abrasives in the engine are dull gray vertical scratches on piston skirts, scratched cylinder bore, high ridge at the top of the cylinder, loose piston fit, or badly scratched rod and main bearings.

Common causes for abrasives in an engine are:

- **No air cleaner**

- **Failure to service air cleaner at regular intervals**

- **Loose connections between air cleaner and intake manifold**

- **Holes in tubing**

- **Damaged filter element or housing, permitting unfiltered air to go around filter element**

- **Failure to properly clean cylinder bores when repairing**

Recommendation: Replace

Fig. 34

Badly worn compression ring is shown at left. A new ring is at right.

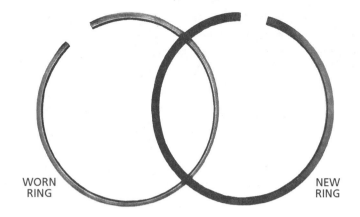

Fig. 35

Vertical scratches across the faces of rings are caused by airborne abrasive or abrasives left in the engine at the time of overhaul.

Unless the source of the abrasives is found and corrected, the life of any new ring set will be short.

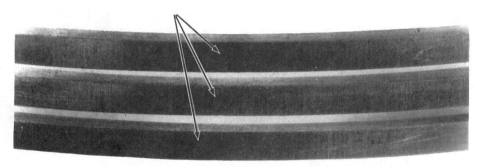

Fig. 36

The wear on this oil ring and steel expander ring was caused by contact with the cylinder wall. The oil ring can no longer provide oil control.

Recommendation: Replace

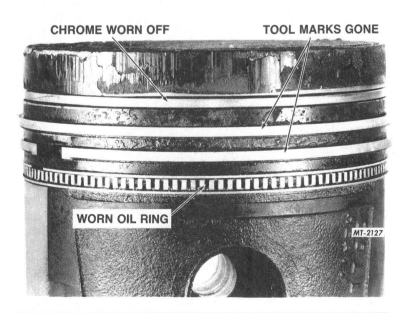

CHROME WORN OFF   TOOL MARKS GONE

WORN OIL RING

MT-2127

Fig. 37

Many aluminum pistons removed for re-ring installation have excessively worn top grooves. This area wears most because it is exposed to maximum combustion heat and pressure and all airborn abrasives that enter the engine. See Technical Manual for specification.

Recommendation: Replace

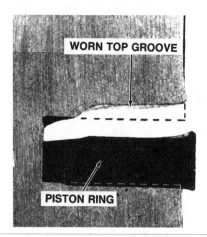

Fig. 38 — Magnified Illustration

CHIPPING

Chrome can be chipped from rings by careless handling or use of incorrect ring compressors. Rings can also be chipped during operation by incomplete combustion.

Recommendation: Replace

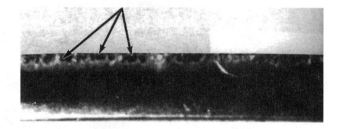

Fig. 39 — Magnified Illustration

SCUFFING AND SCORING

Ring scuffing resembles piston scuffing. Small amounts of ring material have pulled away and stuck to the liner.

Causes of scuffing and scoring are:

- **Overheating due to a faulty cooling system**
- **Lack of cylinder lubrication**
- **Improper combustion**
- **Incorrect or insufficient bearing or piston clearances**
- **Improper break-in**
- **Coolant leakage into cylinder**
- **Overfueling**

Recommendation: Replace

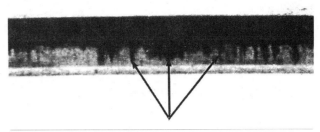

Fig. 40

Scoring is a more severe form of scuffing. Both the rings and the piston are scored. When metal-to-metal contact occurs between the two rubbing surfaces and the temperature of one of these surfaces reaches the welding point of the material, scoring will result.

Recommendation: Replace

Fig. 41

BREAKAGE

When a ring breaks, the pieces bounce around in the groove causing severe erosion of the land and some scuffing on the piston crown or skirt. A broken ring caused erosion of second and third lands which look "melted" away.

Incorrect installation is a major cause of ring breakage. Expanding a ring by hand or using the wrong size expander can fracture a ring causing it to break into pieces during operation.

Ring breakage can also occur when the groove has worn too much or is filled with carbon.

Recommendation: Replace

Fig. 42

STICKING

Deposits caused by too much heat, unburned fuel and excess lubricating oils collect in the piston ring area. Ring failure usually occurs when these deposits harden and freeze the rings in their grooves.

When the rings are completely stuck they often break.

Deposits on the top ring groove cause sticking, scuffing and scoring because they keep out oil and trap metal particles that wear off the piston.

Recommendation: Replace

Fig. 43

Sludge deposits in the oil control ring cause it to plug. This means that oil control has been lost.

Other conditions that lead to stuck or plugged rings are:

- **Plugged air cleaner**
- **Excessive idling**
- **Top groove failure**
- **Cylinder liner distortion**
- **Combustion knock (gasoline)**
- **Preignition (gasoline)**
- **Overloading**
- **Cooling system failure**
- **Improper lube oil**
- **Cold engine operation**
- **Overfueling**

Recommendation: Replace

Fig. 44

IDENTIFYING CYLINDER LINER FAILURES

The principal causes of wet cylinder liner failures are:

- **Cracking**
- **Chemical attack**
- **Erosion**
- **Wear**
- **Scratching**

CRACKING

The failure in the liner consisted of a lengthwise crack originating at the flange and extending downwards below the ring travel area. Apparently water penetrated into the liner, probably as a result of a leaky head gasket and promoted the crack of the liner during compression stroke.

Recommendation: Replace

Fig. 45

The failure in the liner consisted of a crack directly below the flange that penetrated across the section to the top side. The liner could have cracked in overhaul of the engine. The damage may be due to dirt in the counterbore of the block, under the flange, causing uneven loading.

Recommendation: Replace

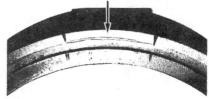

Fig. 46

CHEMICAL ATTACK

The etch marks (arrows) in the ring travel area were caused by the corrosive action of the coolant. The piston is often scored before the corrosion appears on the liner.

Recommendation: Replace

Fig. 47

This apparent crack is actually a line caused by chemical attack.

Recommendation: Replace

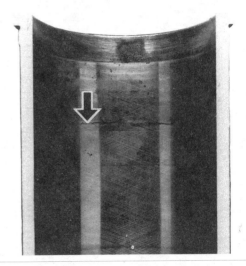

Fig. 48

EROSION

Erosion occurs when bubbles violently collapse against the coolant side of the liner. Such action is accelerated by impurities and lack of proper rust inhibitors in the coolant.

Recommendation: Liner can be reused, but it should be installed 90 degrees from where erosion occurred.

Fig. 49

Liner packing ring failure caused by erosion. Use of conditioner reduces erosion.

Recommendation: Replace

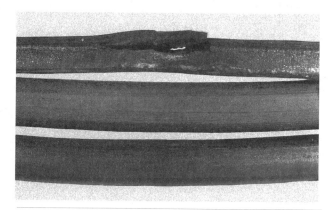

Fig. 50

WEAR

Wear steps are often visible in used liners and normally should be replaced. If clearance between piston and liner is within service specification, and if liner is not otherwise damaged, the liner may be re-useable.

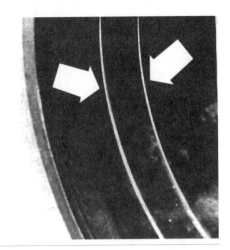

Fig. 51

SCRATCHING

Scratching occurs when dirt particles enter the engine. During operation the dirt moves between the liner and piston rings causing scratches.

Recommendation: Replace

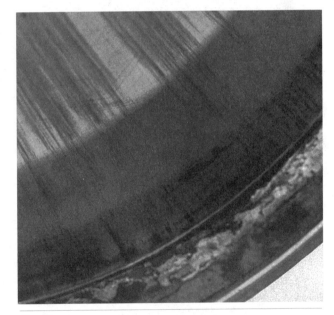

Fig. 52

BLOWN HEAD GASKET

Damage to cylinders, liners, and all engine parts lubricated by oil can be directly and indirectly caused by a blown head gasket. Damaged gaskets can allow coolant leakage into the crankcase of an engine.

Hot gases blowing by the damaged gasket at push rod bores can cause material build up and excessive heat on the push rod.

Push rods can be bent by these added stresses.

Recommendation: Replace

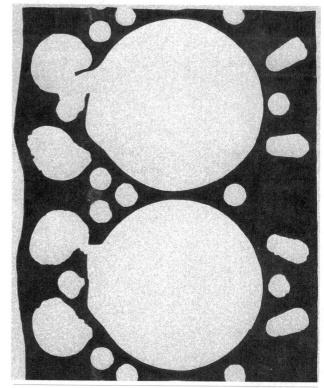

Fig. 53

Similar damage can occur at the cylinder openings.

Gasket damage usually can be traced to incorrectly tightened cylinder head cap screws or faulty gasket material.

Gasket damage at the cylinder bore can occur because of:

- **Faulty material**
- **Incorrect cylinder piston**
- **Incorrect piston ring assembly**
- **Incorrect installation of liner**

Recommendation: Replace

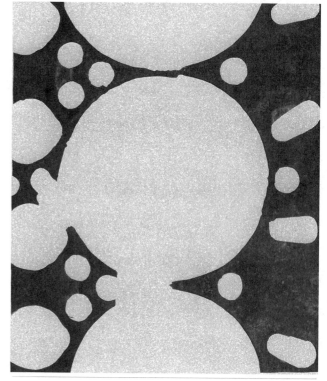

Fig. 54

TEST YOURSELF

QUESTIONS

1. What are three causes of detonation or combustion knock in gasoline engines. (Five causes were given in the text.)

2. Name the condition that occurs when gasoline ignites before the spark is delivered to the cylinders.

3. Name the condition that occurs when two metal parts rub and the resulting heat builds up to the welding point causing small pieces of metal to be pulled from the hotter surface and deposited on the cooler surface?

4. What causes dull gray, vertical scratches on rings and pistons?

5. What causes piston rings to stick?

6. Name one of the three functions of piston rings.

7. Which component is most directly affected by contaminated coolant: a) piston, b) piston rings, or c) cylinder liner?

JOURNAL BEARINGS

2

INTRODUCTION

When replacing damaged bearings it is vital to determine the cause to prevent repeated failure. Most failures are due to the following causes:

- **Dirt**
- **Lack of lubrication**
- **Improper assembly**
- **Misalignment**
- **Overloading**
- **Corrosion**
- **Electric current**

DIRT

Large dirt particles can embed in the soft bearing material. This causes wear and decreases the life of both the bearing and its journal. Dirt is the most frequent cause of bearing failure. Prevent this by cleaning the area surrounding the bearing thoroughly during installation and by proper maintenance of any air and oil filters.

Recommendation: Replace and examine other components for wear or damage

Fig. 1

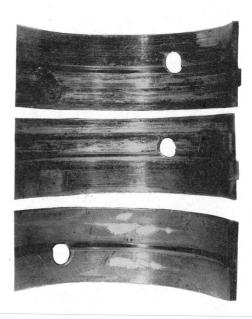

Fig. 2

Fig. 3

This illustration shows a particle of dirt left on the outside surface of the bearing during installation.

Fig. 4

The particle of dirt pushed the bearing inward increasing localized pressure and heat, damaging the inside of the bearing.

Recommendation: Replace

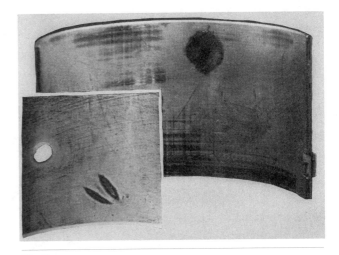

Fig. 5

LACK OF LUBRICATION

Oil starvation damaged these bearings. Lack of oil can occur immediately after overhaul, when priming of the lubricating system is vital.

After break-in, other things can happen. Both local and general oil starvation can result from external leaks. Blocked oil suction screen, oil pump failure, plugged or leaking oil passages, failed relief valve springs, or badly worn bearings can stop the circulation of lubricating oil.

A mislocated oil hole will also cut off the oil supply to a bearing, causing rapid failure. Always check to be sure the oil hole in the bearing is in line with the oil supply hole.

Also, in the case of engines, the oil supply may become diluted by seepage of fuel into the crankcase from a failed fuel pump. This will reduce the oil's film strength and the friction will score the bearings.

Recommendation: Replace

IMPROPER ASSEMBLY

Improper assembly and the resultant damage to the bearing can be due to tapered journals, out-of-round bearing bore, incorrect crush, or rod misalignment.

Tapered journals allow areas of excessive clearance between the journal and bearing, distributing more wear on one edge of the bearing. This wear is increased by the force on the bearing carrying the greater load.

Recommendation: Replace using proper assembly procedures.

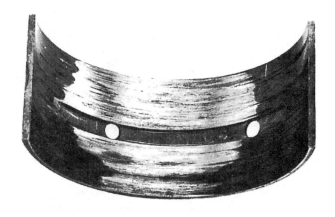

Fig. 6

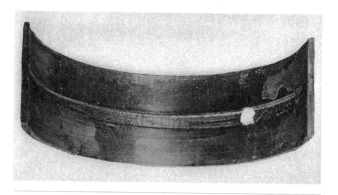

Fig. 7

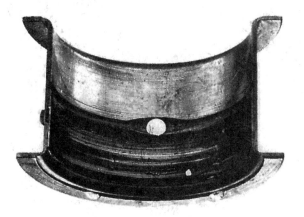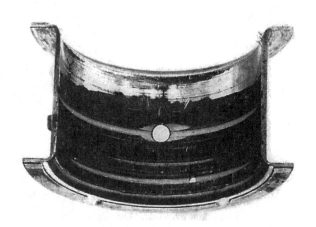

Fig. 8 — Improper Assembly

Out-of-round bearing bore usually is visible by lining wear pattern at parting edges, so high wear rate and heat takes place in this area.

Recommendation: Replace

Fig. 9

A small amount of the bearing insert extends beyond parting edges of rod and rod cap. When the rod bolts are tightened, the bearing inserts are seated. This condition is called "crush." Crush built into the bearing inserts by bearing manufacturers is determined by experience and engineering data, and should under no circumstances be altered. As recommended torque value is applied to the bearing halves, a squeezing action takes place due to correct crush being present at parting edges. This pressure holds the inserts secure in correct position.

Excessive crush frequently results when someone files parting edges of bearing caps. Excessive crush brings about inward collapse of the bearing insert resulting in premature bearing failure and crankshaft damage.

The opposite of excessive crush is insufficient bearing crush which can result in destructive action to both bearing and crankshaft. Any polished area on bearing insert back, or at parting edges, is a true indicator of insufficient crush. Polishing action is caused by bearing movement in bearing bore. Insufficient crush causes loss of heat transfer and bearing lining failure. A few causes of insufficient crush are:

Fig. 10

- **Insufficient torque due to damaged mating surfaces at parting edges of caps and bores**

- **Capscrews bottoming in blind threads, resulting in false torque reading**

- **Bearing bore wear or cap stretch**

Recommendation: Replace

MISALIGNMENT

Heavy wear on outer edges of upper and lower bearing inserts could indicate misaligned connecting rod.

Misalignment is caused by:

- **Operational abuse such as excessive lugging**
- **Faulty connecting rod installation**
- **Abuse of connecting rod in work area prior to assembly in engine**

Recommendation: Replace

Fig. 11

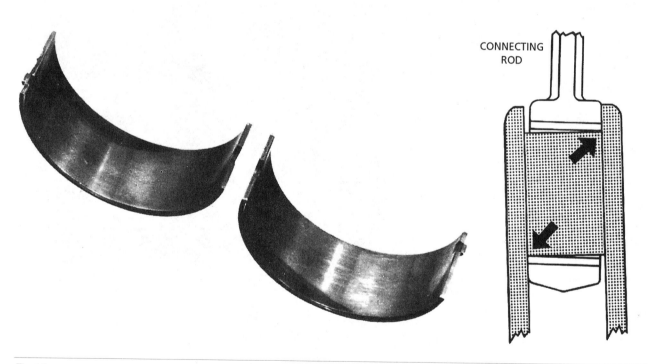

CONNECTING ROD

Fig. 12

Misalignment can cause concentrated wear on the bearings — one edge of the upper bearing and the opposite edge of the lower bearing. When this wear pattern exists, check the alignment of the shaft and bearings.

Recommendation: Replace

OVERLOADING

Overheating from overloads causes a metal fatigue which breaks away from the surface of the bearing.

Recommendation: Replace

Fig. 13

CORROSION

Corrosion from acid formation in the oil causes finely pitted surfaces and large areas of deterioration.

Corrosion occurs when oil temperature goes very high and when excessive blow-by occurs in engines. Condensation, and in some cases, incorrect lubricant will also cause corrosion.

Prevent corrosion in engine by following the manufacturer's recommendations for:

- **Oil viscosity and service grade**
- **Oil change intervals**
- **Coolant and coolant change intervals**

Recommendation: Replace

Fig. 14

This shows corrosion of the lead in a copper-lead bearing. Probably the biggest source of corrosive attack is the oxidation products formed in oil itself.

Recommendation: Replace

Fig. 15

Cavitation erosion is a mechanical washing away of the bearing surface between the bearing and the journal.

Heating and engine vibration can form air bubbles in the oil. The collapse of the air bubbles can cause extremely high localized pressure which results in fatigue and pitting of the bearing surface.

Recommendation: Replace

Fig. 16

ELECTRIC CURRENT

Destructive current may cause sparks to pass through the oil film. This, in turn, produces microscopic pits in the surface of the journal and the bearing. This action results in a continual removal of metal from the surface of the bearing. Troubles may be noticed in periods ranging from a few hours up to several years, depending upon the magnitude of the current flow.

Electrical wear shows up as a frosted surface in the loaded zone of the bearing.

Recommendation: Replace

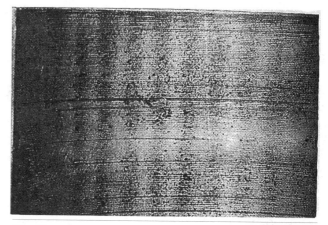

Fig. 17

TEST YOURSELF

QUESTIONS

1. What is the most frequent cause of journal bearing failure?

2. How can we prevent the most frequent cause of journal bearing failure?

3. What is essential to the seating (or "crunching") of the bearing inserts?

4. What does concentrated wear on one edge of the upper bearing and the opposite edge of the lower bearing indicate?

5. What does finely pitted surfaces with large areas of deterioration on the bearings indicate?

VALVE GEAR TRAIN

3

INTRODUCTION

The valve gear train in a typical engine consists of the camshaft, tappets, push rods, rocker arms, valves and valve springs. The failure of these parts are described in this section.

IDENTIFYING VALVE FAILURES

Of all the components of the valve gear train, valve failures, particularly exhaust valves, represent the more common problems.

The major causes of valve failures are:

- **Distortion of the valve seat**
- **Deposits on valve**
- **Too little tappet clearance**
- **Scuffing of stem**
- **Erosion**
- **Heat fatigue**
- **Pitting**
- **Breaks**
- **Wear**

DISTORTION OF VALVE SEAT

The valve is burned because of distortion of the valve seat. Major causes of seat distortion are:

- **Failure in the cooling system.**

- **Out-of-round or loose seat. This may stop the transfer of heat between the insert and the head or block.**

- **Warped sealing surfaces on heads or blocks often distorts the seats when the head is tightened. Improper tightening, too much torque and the wrong sequence can also distort valve seats.**

- **Failure to grind the valve seat concentric with the valve guide bore.**

Recommendation: Replace

Fig. 1

Intake valves "cup" (A) when they are used in an exhaust port due to excessive stress caused by overheating.

Intake valve stem discoloration (B) indicates too much heat.

Recommendation: Replace

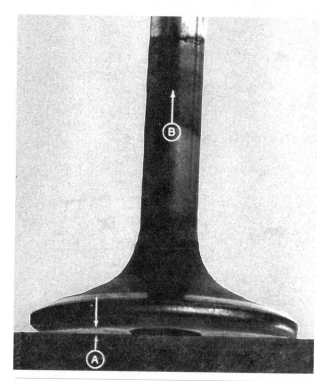

Fig. 2

DEPOSITS ON VALVE

Deposits are not typical causes of failure in diesel engines, but do occur.

Deposits on the valve stem can be the result of high temperatures caused by using an improper oil.

There was no heat dissipation from the valve stem to the valve guide and from the guide to the cylinder block.

The heat was trapped in the valve stem or valve guide causing excessive carbon formation.

Recommendation: Replace

This valve failed because face deposits built up and then broke off. This damaged the seat and the resulting "blow-by" burned the valve.

Fig. 3

Other factors that may cause this type of failure are:

- **Weak valve spring causes a poor seal between the seat and the face, allowing deposits to form.**
- **Too little tappet clearance also causes a poor valve-to-seat seal.**
- **Valves sticking in the valve guide allows deposits to build up on the valve face and seat.**
- **Valve seats that are too wide cut down the seating pressure and reduce the crushing of deposits when the valve closes.**
- **Lack of valve rotation (which is needed to "scrub" the valve).**

Recommendation: Replace

Fig. 4

TOO LITTLE TAPPET CLEARANCE

Failure of the valve was caused by too little tappet clearance. The valve was held off its seat and blow-by caused face-burning. Causes of too little tappet clearance are:

- **Tappet clearance not set to specifications**
- **Failed valve rotators and, as a result, burned valves**
- **Cooling system not operating properly or wrong thermostat**
- **Extremely high temperatures affect tappet clearance**
- **Tappet clearance not rechecked after break-in and retightening of head**

Recommendation: Replace

Fig. 5

BURNED VALVE

The valve burned and failed as a result of preignition. Valve temperature became so high that part of valve face melted away.

Recommendation: Replace

Fig. 6

EROSION OF VALVES

This valve is eroded, but has not failed. However, it would have broken after much more service because of the erosion under the head.

Causes of valve erosion are:

- **Wrong type of fuel**
- **Faulty combustion**
- **Too high valve temperatures**
- **Lean fuel air mixtures which overheat valves and erode them**

Recommendation: Replace

Fig. 7

HEAT FATIGUE

Overheating can crack the valve head. More cracking may cause parts of the valve to break off.

Causes of cracked valves from heat fatigue are:

- **Worn guides**
- **Distorted seats**
- **Lean fuel-air mixtures**

Recommendation: Replace

Fig. 8

Excessive heat warped this exhaust valve stem and damaged the seal. The heat was created by overspeeding when a malfunctioning governor overfueled the engine.

Recommendation: Replace

Fig. 9

PITTING

Carbon particles can build up between the valve and the valve seat. This can cause pitting on the valve surface.

Recommendation: Replace pitted valves and regrind the valve seats.

Fig. 10

BREAKS

Fatigue break is the gradual breakdown of the valve due to high heat and pressure. A fatigue break usually shows lines of progression as at the top.

Impact break is the mechanical breakage of the valve. The cause is seating the valve with too much force, often caused by too much valve clearance. An impact break does not show the lines of progression, but rather the familiar crow's-feet (see bottom).

Broken valves are not always clearly one type or the other. Combinations of heat and high seating force can produce failures of varying degrees and appearance.

Recommendation: Replace

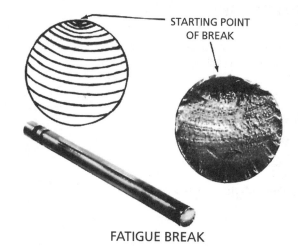

STARTING POINT OF BREAK

FATIGUE BREAK

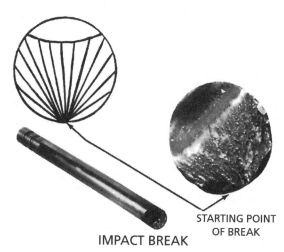

STARTING POINT OF BREAK

IMPACT BREAK

Fig. 11

The valve head (bold arrow) in the top of this piston is the result of a mechanical break of the valve stem caused by overspeeding of the engine. A defective seal in the turbo-charger allowed engine oil to leak through the intake manifold into the cylinder combustion chamber. The overspeeding occurred as the oil ignited.

Recommendation: Replace

Fig. 12

WEAR

This valve failed because of face burning. Wear on the stem and carbon projecting into the guide shows the valve guide was worn and was the likely cause of the failure.

Worn valve guides lead to valve failures:

- **Worn guides prevent even grinding of the valve seat leading to out-of-square seating, which allows burning gas to leak out and burn the valves.**

- **Worn guides cause the valves to strike at an angle and damage the sealing surface leading to blow-by and burning.**

- **Excessive stem-to-valve guide clearance allows too much oil to run down the valve stem causing excessive carbon deposits resulting in stuck rings and/or poor heat dissipation.**

- **When the inside edges of the valve guides wear, they can no longer act as carbon scrapers.**

Other factors can also cause valve guides to fail prematurely:

- **Worn rocker arms cause excessive side thrust on the valve stem.**

- **Poor lubrication results in scoring.**

- **Carbon deposits on the valve stem wear the valve guide into a bell-mouthed shape.**

- **Cocked valve spring places side thrust on the valve stem, and results in excessive wear.**

Fig. 13

IDENTIFYING ROCKER ARM FAILURES

An example of poor lubrication on rocker arm tip wear is shown. The rocker arm tip in this case was worn approximately 1/32" deep. The effect of this condition would:

- **Upset valve motion.**
- **Contribute to side thrust of the valve, thus aggravating guide wear.**
- **Make lash adjustment with a conventional feeler gauge difficult.**

Recommendation: Replace

Fig. 14

Spalled rocker arm inserts are shown. Spalling results from heavy loading and repeated contact between moving parts.

Recommendation: Replace

Fig. 15

Loose rocker arm bracket bolts allowed the rocker arm shaft and bracket to bend and break. Whenever the rocker arm shaft is installed, the mounting bolts should be tightened to the specifications in the technical manuals. Scuffed rocker arm balls at the push rod and rocker arm tips also result in failures.

Recommendation: Replace

Fig. 16

IDENTIFYING PUSH ROD FAILURES

When a push rod cup is worn or broken it also can cause a valve failure. The rod has a broken cup caused by misalignment of the push rod with rocker arm adjusting screw. Careful reinstallation of the rocker arm assembly prevents this.

Recommendation: Replace

Fig. 17

BENT PUSH ROD

Bent push rods are often caused by a blown head gasket that allows blowby of hot gases on the rod.

Recommendation: Replace

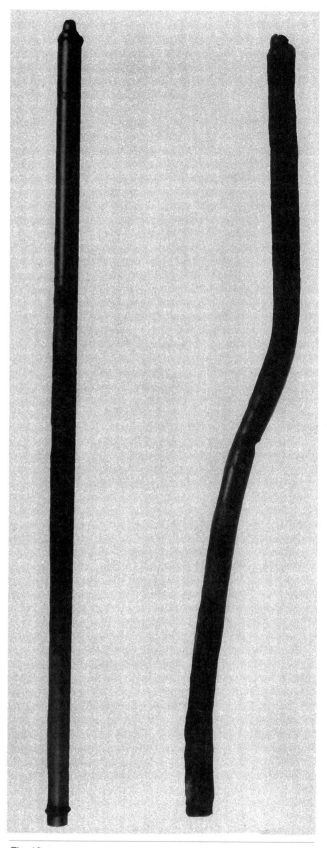

Fig. 18

IDENTIFYING VALVE TAPPET FAILURES

Because of heavy loads and high speeds in engine operation, valve tappets may show signs of wear, scuffing and surface fatigue. Wear is a gradual, smooth loss of surface material by friction (left). Scuffing is a more severe form of wear (center). The deterioration is characterized by deep grooves, rapid surface roughening, welding and tearing away of the surface metal. Surface fatigue is a tendency of a metal chip to break under stress conditions (right). It is frequently referred to as pitting, spalling and flaking. Dust in the oil causes abrasive wear.

Recommendation: Replace

Fig. 19

TEST YOURSELF

QUESTIONS

1. Which component of the valve gear train is most likely to fail?

2. What causes discoloration of intake valve stems?

3. What causes excessive carbon deposits on valve stems?

4. What can cause burning of the valve face?

5. (True or false?) Preignition can cause temperature in a valve to rise high enough to melt the valve face.

6. (True or false?) The wrong type of fuel or a lean fuel air mixture can cause erosion of the valve head.

7. What causes cracks in the valve head?

8. What causes pitting of the valve surface?

9. What causes an "impact" break in a valve?

10. (True or false?) Worn valve guides do not seriously affect valve operation.

TURBOCHARGERS

4

INTRODUCTION

This section deals with identifying failures of the following turbocharger components:

- **Wheels and Impellers**
- **Shafts**
- **Journal Bearings**
- **Thrust Bearings**
- **Housings**

These failures are discussed and illustrated on the following pages.

IMPORTANT: In almost all cases, replacing a failed part will not, in itself, correct the failure. To avoid repeated failures, determine and correct the cause of failure.

WHEELS AND IMPELLERS

High rotational speeds and high temperatures make the turbocharger wheels sensitive to abusive environmental and operating conditions.

The causes of turbocharger wheel failures are:

- **Foreign material (a major cause of failure)**
- **Contact damage**
- **Erosion or abrasion**

FOREIGN MATERIAL

Foreign material is a major cause of turbocharger failure. Foreign material in the intake system may lead to this type of damage. Badly nicked edges of the compressor wheels were probably caused by:

- **Welding slag not removed from duct**

- **Loose wire particles from the air cleaner**

- **Nuts, bolts, washers, etc.**

- **Particles left from previous turbocharger failures**

- **Compressor wheel nut loosened and backed off threaded shaft**

Recommendation: Replace

Fig. 1

Foreign material in the exhaust system damaged the turbine wheels. Damage is fairly uniform on all blades. Probable causes of blade tips chewed, battered, deformed and broken off are:

- **Loose material in exhaust manifold (nuts, bolts, washers and pieces from previous turbocharger failure)**

- **Engine valve breakage**

- **Piston ring breakage**

Recommendation: Replace

Fig. 2

Fig. 3

CONTACT DAMAGE

Damage to bearings from contaminated lubricant or from lack of lubricant will allow motion of the shaft. This may permit the compressor wheel or turbine wheel to contact their respective housings. This could also be caused by an imbalanced rotating assembly.

Finding more than one blade broken with heavy rubbing on adjacent blades gives evidence that blade breakage followed, rather than caused, the contact damage.

Recommendation: Replace

Fig. 4

EROSION OR ABRASION

Sand caused this impeller erosion. Note the nicked leading edges, blade tips thinned from sand or other hard particles striking the blade surface at a high rate of speed, and blade contour deeply eroded.

Recommendation: Replace

Fig. 5

SHAFTS

The modes and causes of turbocharger shaft failures are described below:

- **Sludge build up**

- **Oil starvation (a major cause of turbocharger failure). If an engine is shut off immediately, without cooling down, the turbocharger "free-wheels" to a stop and receives no lubrication. Allow the engine to run for a short time at low idle before shut off**

- **Contaminated oil**

SLUDGE BUILD UP

The probable cause of sludge build up on the turbocharger shaft and bearing (A) is the shutdown of the engine immediately after full load operation without allowing the engine to cool down properly. Excessive heat can result. This also can discolor the shaft and bearing (B). Overheated bearings will close off oil holes. This will lead to further lubrication problems.

Sludge buildup can result in:

- **Turbine and compressor wheel rub**

- **Shaft damage/breakage**

- **Thrust washer wear or damage**

Recommendation: Replace

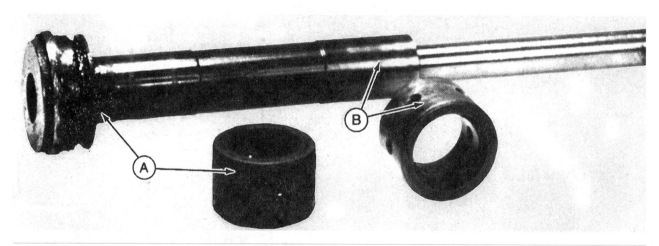

Fig. 6

OIL STARVATION

Start-up and shut down procedure is important especially after an engine has not been used for several weeks, or after an oil change. Momentary oil starvation can cause bearing material to rub off (arrow), on the shaft and cause shaft vibration.

Recommendation: Replace

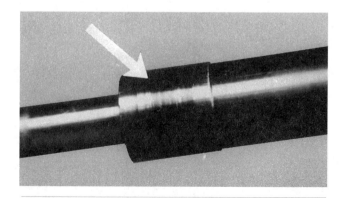

Fig. 7

A momentary loss of oil can also cause blueing of the shaft, a sign of overheating.

Recommendation: Replace

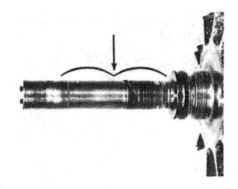

Fig. 8

CONTAMINATED OIL

Contaminated oil can damage internal parts of a turbocharger. Scratched and worn bearing surfaces will cause drag and unbalance the rotating assembly resulting in turbocharger failure.

Recommendation: Replace

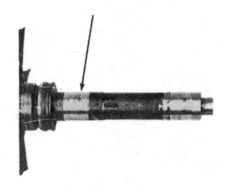

Fig. 9

JOURNAL BEARINGS

The principal causes of journal bearing failures are:

- **Oil starvation - (Note—a major cause of turbocharger failure). Engines must be run for a short time at low idle speed before being shut off. This allows turbocharger and engine components to cool down before the oil supply is cut off**

- **Contaminated oil**

OIL STARVATION

Lack of lubricant caused the bearing material to deform (oil holes beginning to fill in).

Fig. 10

Inadequate lubrication caused the bearing to look spotty with fine scratches.

Recommendation: Replace

Fig. 11

Bronze journal bearings will lose their tin plating and discolor when overheated. The plating (left) is worn and missing in spots. A new bearing is at right.

Recommendation: Replace

Fig. 12

CONTAMINATED OIL

Grooves in the journal bearing may appear and heavy, deep scratches on the outer diameter of journal bearings indicate abrasive matter in the lubricating oil.

Fig. 13 — Grooves in Journal Bearing

The bearing on the right is normal wear.

Recommendation: Replace

Fig. 14 — Scratches in Journal Bearing

Outer bearing surface is scored and worn and the plating missing because the lubricant was contaminated.

Recommendation: Replace

Fig. 15

Contaminated oil caused the bearing and shaft scoring.

Recommendation: Replace

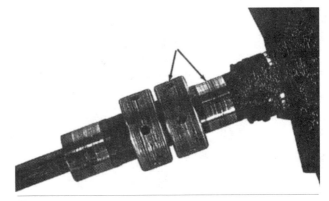

Fig. 16

Contaminant may become imbedded in aluminum bearing surfaces and cause heavy wear on shaft journals and bearings.

Recommendation: Replace

Fig. 17

UNBALANCED ROTATING ASSEMBLY

An unbalanced rotating assembly has a tendency to pound the bearing. This will generally shrink the size of the oil passages, reducing the oil supply.

Recommendation: Replace

Fig. 18

The bearing can seize in the housing bore if unbalance is not corrected. The shaft will continue to pound the bearing until the oil passages are closed.

Recommendation: Remove and replace seized bearings.

Fig. 19

THRUST BEARINGS

The principal causes of thrust bearing parts failures are:

- **Oil starvation (a major cause of turbocharger failure)**
- **Contaminated oil**

OIL STARVATION

Heat discoloration of thrust rings indicates a lack of lubricant. Often rubbing marks are present (arrow).

Recommendation: Replace

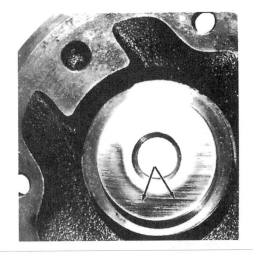

Fig. 20

CONTAMINATED OIL

Foreign material in the oil causes erosion around the oil passages.

Recommendation: Replace

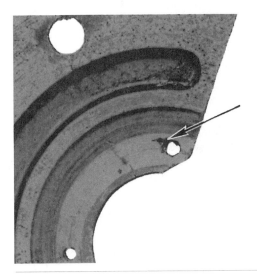

Fig. 21

Contaminated oil can also wear the thrust collar. Excessive end play can develop.

Recommendation: Replace

Fig. 22

Shaft motion and bearing damage caused by lack of lubricant or foreign material can, in turn, break thrust bearing parts. Thrust washer worn and broken away is shown.

Recommendation: Replace

Fig. 23

HOUSINGS

Turbine housing cracks can result from foreign material entering the housing, excessive temperature, or thermal stress. In some instances it appears as a slow wearing away of the housing. Eventually it terminates in rupture.

Recommendation: Replace

Fig. 24

Extreme temperature cracks are identified as random hairline cracking. High temperatures can be caused by exhaust and intake leaks, gross overfueling, use of the wrong turbocharger for the engine, and abuse of the turbocharger, especially at high altitude conditions.

Recommendation: Replace

Fig. 25

If the seal ring area of the housing is heavily rubbed by the turbine wheel hub, the cause is often shaft motion or journal bearing damage.

Recommendation: Replace

Fig. 26

TEST YOURSELF

QUESTIONS

1. What is one cause of turbocharger wheel failure? (Three were given in the text.)

2. Why is it important to idle the engine a short while before shutting it off?

3. What causes grooves, scratches, and scoring in the turbocharger journal bearings?

4. What causes random hairline cracking of the housing?

5. (True or false?) Replacing the failed part will usually prevent future failures.

6. What are the two major causes of turbocharger failure?

GEARS

INTRODUCTION

The principal causes of gear failure are:

- **Wear**
- **Pitting, spalling, case crushing**
- **Fatigue**
- **Impact**
- **Rippling, ridging and cold flow**
- **Combined effects**

Many gear failures result from overloading a gear or from impact and shock loads brought on by improper shifting or clutching. Under normal recommended loads, most gears perform satisfactorily.

New gear teeth usually have slight imperfections that normally disappear during break-in as the teeth are oiled and polished. After break-in the teeth should have a long service life if they are properly lubricated, operated, and adjusted.

Only an examination by a metallurgical laboratory can determine if manufacturing imperfections exist in a gear.

The following sections on gears identify:

- **General category failures – common to all gears, regardless of application and assembly conditions**
- **Specific category failures – common to gears in rear axles and transmissions**

GEAR TERMINOLOGY

Parts of the gear that are discussed in this chapter are shown in the illustration to the right. These terms apply to all types of gears.

Some gears are case-hardened by heat treatment after they are machined to the final form. A hardened case (A) is produced by heat treatment providing a hard, wear resistant outer surface supported by a lower-hardness, tough core.

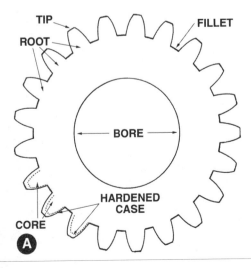

Fig. 1

Backlash is the "play" or clearance of two gears in mesh. Too much backlash can cause severe impact on the gear teeth from sudden stops or reverses of the gears. Too little backlash can cause overload wear on the teeth, premature gear failure, and excessive stress on shafts and bearings.

Consult the service or technical manual for the individual machine for the proper backlash.

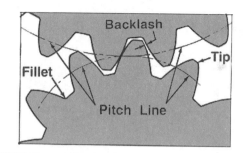

Fig. 2

GENERAL CATEGORY OF FAILURES

WEAR

Wear is a removal of surface material from the gear. It can be slow, such as scuffing, or rapid, such as scoring. There are three types of wear;

- **Adhesive wear - caused by metal-to-metal contact with surfaces welding together, then tearing apart. Possible causes are inadequate lubrication or the gears not properly in mesh.**

- **Abrasive wear - caused by foreign particles, such as dirt and grit.**

- **Corrosive wear - a chemical attack of the gear surface from contaminated oil or an additive.**

An adhesive type of wear is shown. Possible causes are inadequate lubrication or improper gear mesh.

Recommendation: Replace

Fig. 3

Moderate wear on the face of gear teeth, causes the operating pitch line to become visible (arrow). The wear probably was caused by abrasive material in the lube oil.

Recommendation: Replace

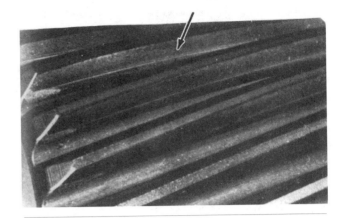

Fig. 4

The gear suffered scoring because of metal-to-metal contact under heavy pressure, resulting from inadequate lubrication. The horizontal line on the worn surface indicates the pitch line (arrows).

Recommendation: Replace

Fig. 5

The early stages of scoring shows a spotty frosting pattern on the upper portion of the teeth. Damage is slight at this stage.

Recommendation: Check for proper lubrication and gear mesh. Reuse.

Fig. 6

Destructive scoring is shown. Heavy scoring occurred above and below the pitch line. Usually, the damage progresses quickly and the gear fails.

Recommendation: Replace

Fig. 7

An abrasive type of wear is shown. Dirty lubricant or not following the recommended service intervals causes abrasive wear.

Recommendation: Replace

Fig. 8

A particularly severe form is shown. A large portion of the tooth of the sintered-metal or bronze pinion has worn away due to an accumulation of abrasive particles in the lubricant.

Recommendation: Replace

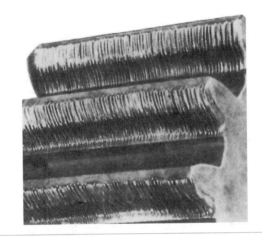

Fig. 9

Corrosive wear is shown. It was caused by contaminants or additives in oil.

Recommendation: Replace

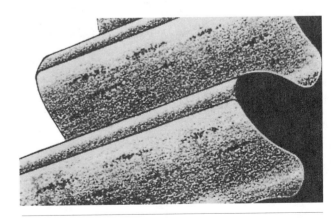

Fig. 10

This gear surface was damaged by a chemical reaction. This kind of wear will continue until the gear fails. Chemical wear results from contaminated oil, composition of oil, or an additive.

Recommendation: Replace

Fig. 11

PITTING, SPALLING AND CASE CRUSHING

Pitting is a type of fatigue failure. Small particles of the gear separate from the tooth surface. Excessive stress on the surfaces of the mating gears causes the pits or separation of the particles. Overloading the gears creates the stress.

The pits normally start along the line of contact where the pressure is heaviest against the teeth of the mating gears. A fatigue crack will frequently start at a pitted area.

Spalling is an advanced state or severe form of pitting. Part of the gear may crack off when this condition is reached.

Case crushing is the crushing of the outer, hardened surface of the gear teeth. Cracks running along the face of the teeth is an indication of case crushing. Case crushing, like pitting and spalling, is usually caused by excessive operating loads being exerted on the gears.

The "corrective" pitting in the hypoid pinion consists of very small pits that do not progress beyond the initial stage, and frequently "heal over."

Recommendation: Reuse

Fig. 12

This pitting began on the outside end of the helix (the right edge of the gear in this picture), because of a small amount of misalignment, and worked its way to the middle of the tooth. Eventually pitting stopped and the surface began to polish, indicating that the load across the teeth became more evenly distributed. This type of pitting is not harmful.

Recommendation: Reuse

Fig. 13

In contrast, this destructive pitting, probably resulted from excessive loads.

Recommendation: Replace

Fig. 14

Pitting destroyed the surfaces of the teeth in this spur gear.

Recommendation: Replace

Fig. 15

Pitting occurred at the heel contact. Heavy contact occurred at the location of the pits because the tooth surfaces did not mate properly, possibly because of an overload.

Recommendation: Replace

Fig. 16

In the first stage of spalling, cracks running lengthwise developed on the contact area of the teeth shown at left. Spalling completely destroyed the gear when the large fragments came out, as shown at right.

Recommendation: Replace

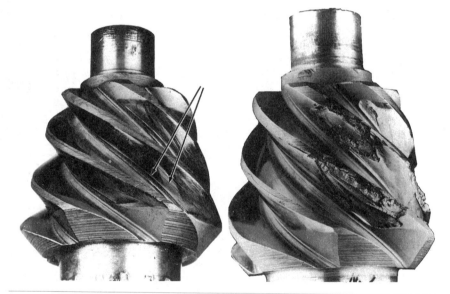

Fig. 17

This drive pinion shows severe heavy pitting, spalling, and complete tooth destruction on successive teeth.

Recommendation: Replace

Fig. 18

G e a r s

Case (hardened surface) crushing, is shown by lengthwise cracks in the contact surface of this bevel gear. The major cracks began deep in the case-core structure and worked their way to the surface. Long chunks of material appear about to break loose from the surface.

Recommendation: Replace

Fig. 19

The initial appearance of case crushing of a carburized gear (a gear hardened by heating the surface, converting it to high-carbon steel, then quenching) is shown at the left and the final stages at the right. The gear probably was overloaded but a metallurgical analysis may be necessary to determine the possible causes of failure.

Recommendation: Replace

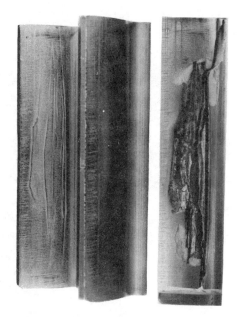

Fig. 20

FATIGUE

Fatigue is generally due to repeated, excessive loads that fracture gear teeth, usually at or near the tooth root.

Fatigue failure may start as a small crack, caused by an extremely high load, and continue under normal operation until the gear fails. The fractured face usually consists of two zones:

- **The smooth fatigue zone, with progressive stages of crack growth.**

- **The final-fracture zone which is rough.**

A typical fatigue failure of a ring gear, with its characteristic smooth zone, is shown. The failure probably resulted from a heavy load or shock load (improper shifting or clutching).

Recommendation: Replace

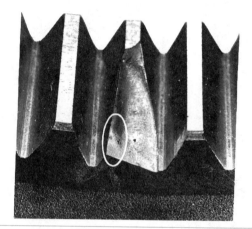

Fig. 21

Fatigue is shown at the root of three pinion teeth. The remaining teeth show only light wear.

Recommendation: Replace

Fig. 22

This oil pump idler gear failed from fatigue with severe tooth wear. The fracture extends from root fillets (the area connecting the bases of the teeth) to the bore in the gear. The gear should be examined by a metallurgical laboratory to determine the cause of failure.

Recommendation: Replace

Fig. 23

This fatigue crack extends from the root fillet to the bore. A metallurgical laboratory examination would be required to determine the possible causes of failure.

Recommendation: Replace

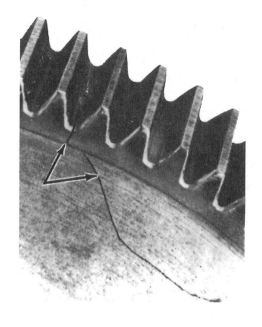

Fig. 24

Fractures caused by repeated, heavy loads on several teeth of a spur gear are shown. The tooth marked "A" apparently broke first (note smooth, velvety zone) as a result of a fatigue crack.

Recommendation: Replace

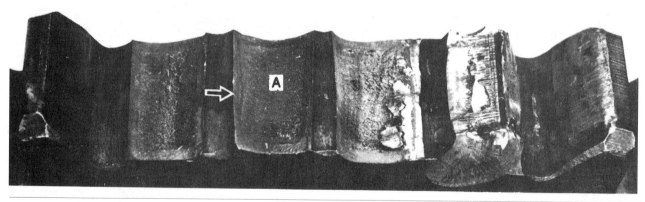

Fig. 25

Another fatigue fracture is shown. The arrow indicates the crack origin at the left edge of the fracture, where a small pit existed near the base of the contact area on the pressure side of the tooth. The area just beside the crack origin (at arrow) is well rubbed, which indicates that the crack progressed slowly at first.

Recommendation: Replace

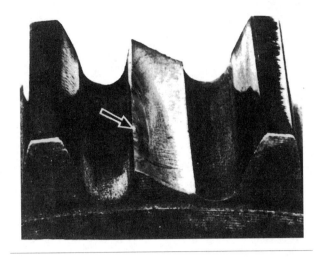

Fig. 26

This gear failure consists of one broken tooth and cracks in the adjacent teeth. Since the fracture face was not smooth it indicates an instantaneous fatigue fracture. A pronounced band of inclusions (impurities in the gear), which act as notches promoting failure, were also observed.

Recommendation: Replace

Fig. 27

Fatigue cracks in a case hardened gear (hardening the outside surface of the gear) began in tooth roots, forming on either side of a tooth (arrow), and meeting in the middle of the tooth. A metallurgical examination is probably required to determine the cause of failure.

Recommendation: Replace

Fig. 28

The planet gear had most of the tips of the teeth broken off. Overloading first caused stresses which initiated cracks on the teeth. These cracks continued, probably by fatigue, to the surfaces of the teeth. If this condition occurs with a relatively new gear, the case hardening treatment may have gone too deep. An examination by a metallurgical laboratory is needed to make this determination.

Recommendation: Replace

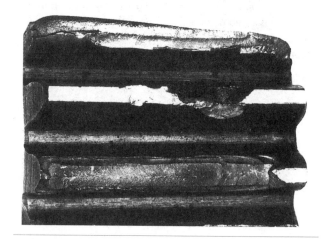

Fig. 29

IMPACT

Impact failures are generally caused by severe loads resulting from abusive service. The failure usually takes place at or near the tooth root and the fractured face is gray and grainy, without evidence of progressive failure.

The battered and chipped tooth corners of this carburized gear show they received repeated blows before fracturing. Improper shifting probably caused this failure.

Recommendation: Replace

Fig. 30

The break in this hardened spur gear has a gray, grainy appearance typical of a shock failure. It does not have the smooth appearance characteristic of fatigue. The failure probably resulted from improper clutching or improper shifting.

Recommendation: Replace

Fig. 31

RIPPLING, RIDGING, AND COLD FLOW

Rippling, ridging, and cold flow (metal movement under high pressure at room temperature) occur less frequently than those previously described and tend to be less damaging.

The surface of this gear shows waviness; a typical case of rippling on a hardened hypoid pinion. Rippling generally occurs on highly loaded gears.

Recommendation: Replace

Fig. 32

This hardened gear shows ridging possibly caused by overloading.

Recommendation: Replace

Fig. 33

This shows an advanced stage of cold flow in a medium hard gear. Such gears have a greater tendency for cold flow than case hardened gears. Material has been rolled over the top edges of the gear teeth resulting in a destruction of the gear tooth profile. Heavy loads probably caused this type of metal movement.

Recommendation: Replace

Fig. 34

This medium hard gear shows surface deformation due to rolling and hammering action. This gear probably was overloaded and operated long after the initial damage occurred, resulting in a battered surface.

Recommendation: Replace

Fig. 35

COMBINED EFFECTS

Severe contact stresses caused plastic flow at the surface of this gear producing ripples. Spalling (chipping of surface material) also occurred near the tooth center. Excessive loading may be the cause of this type of failure.

Recommendation: Replace

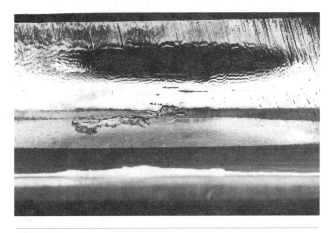
Fig. 36

This carburized drive pinion is pitted near the pitch line. Also apparent are rippling along the pitch line and slight adhesive wear near the top, probably caused by overloading the gear.

Recommendation: Replace

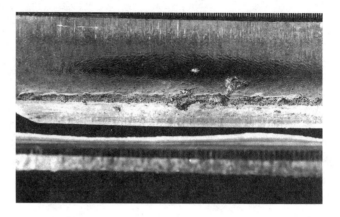

Fig. 37

This final drive pinion shows a complete destruction of the contact portion of the pinion teeth. This was probably caused by heavy loading or inadequate lubrication. The gear teeth extended to the end of the shaft before the failure occurred.

Recommendation: Replace

Fig. 38

These pinions show advanced pitting and adhesive wear. Extreme overloading and inadequate or improper lubrication were probably the causes of these failures.

Recommendation: Replace

Fig. 39

These gears show severe battering and fatigue, probably caused by severe overloading or inadequate lubrication.

Recommendation: Replace

Fig. 40

SPECIFIC CATEGORY OF FAILURES

RING GEAR TEETH

A typical example of a fractured bevel gear tooth caused by improper adjustment is shown. The failure resulted from excessive loading at the heel section of the bevel gear.

Excessive backlash probably caused the failure.

Recommendation: Replace

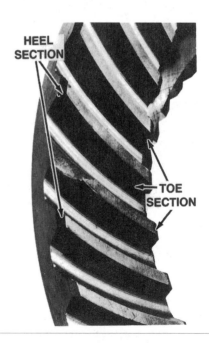

Fig. 41

This fracture resulted from excessive loading on the toe section of the gear, while the cause of failure was insufficient backlash.

Shock loading may also cause these conditions; even to the extent of breaking entire ring gear teeth.

Recommendation: Replace

Fig. 42

RING GEAR AND DRIVE PINION

This gear shows scuffing over the tooth area. The metal had softened and was drawn across the tooth face.

Abnormal friction between gears creates heat which softens the metal and damages the teeth, if not adequately lubricated.

Worn pinion bearings permit end play by the pinion, causing incorrect tooth contact between pinion and ring gear.

Excessive torque could also be a cause of failure.

Recommendation: Replace

Fig. 43

These ring gear teeth show discoloration and distortion. This type of failure is caused by heat generated from improper lubricant, low lubricant level, or infrequent lubricant change. In the presence of any of these conditions the surface overheats because of excessive friction.

Recommendation: Replace

Fig. 44

DRIVE PINIONS

Excessive loading during severe service resulted in concentration of pitted areas at tooth heel. Under excessive loading, deflection throws the pinion out of correct position in relation to ring gear and concentrates the load on the tooth heels.

The smooth zone on the face of the broken teeth is where the crack started. The rough zone indicates an area insufficient to resist a single load application.

Severe operation, defective material, and inadequate radius at the root of the tooth or poor bonding of the radius and the tooth face can cause this type of failure.

Recommendation: Replace

Fig. 45

Fig. 46

DIFFERENTIAL SIDE GEARS AND PINIONS

The grainy, gray appearance of these broken teeth denotes shock failure.

Abusive operation of the machine, such as excessive clutching, creates stresses greater than the maximum strength of the gears.

Recommendation: Replace

Fig. 47

DIFFERENTIAL SPIDER ARMS AND SIDE PINIONS

These spider arms and side pinions are discolored by heat. Evidence of metal to metal contact, scoring and seizing.

This kind of damage is caused by inadequate lubrication, wheel spinning or an overload.

Any of the three will cause the loss of lubricating film between the mating surfaces. Without the oil protection, friction causes the contacting areas to overheat, score, and finally seize if running is continued.

Recommendation: Replace

Fig. 48

REAR AXLE DIFFERENTIAL SIDE GEARS

The small smooth zone of the face of fracture is where this crack started. The rough fracture zone is the result of subsequent failure.

Abusive operation or loose wheel bearings caused a misalignment in axle shaft and housing.

Recommendation: Replace

Fig. 49

TRANSMISSION HELICAL GEAR

The grainy gray appearance of this fracture, with no progressive marks is characteristic of fatigue.

Abusive operation results in extreme shock loads. This failure is frequently associated with speeding up the engine of a load-stalled vehicle and suddenly engaging the clutch.

Recommendation: Replace

Fig. 50

TRANSMISSION SPLINED MAINSHAFT

Coasting a loaded vehicle with clutch disengaged and transmission in gear leads to this type of fatigue failure with a grainy, gray appearance. When the clutch is engaged, the sudden deceleration will produce shock loads and damage various units of the drive line, in this case the mainshaft.

Recommendation: Replace

Fig. 51

TRANSMISSION COUNTERSHAFT

These snubbed, chipped and cracked engaging teeth indicate careless shifting and failure to synchronize the speed of the two gears. Assure proper engagement and synchronization of gears.

Recommendation: Replace

Fig. 52

TRANSMISSION MAINSHAFT SECOND SPEED GEAR

These gear teeth chipped and broken away from the engaging ends of the teeth indicate partial engagement and dragging clutch, which keeps the countershaft turning so that there is interference with gear engagement.

Recommendation: Replace

Fig. 53

TRANSMISSION MAINSHAFT OVERDRIVE GEAR

Lugging the engine under low speed and maximum torque caused extremely high pressures at the surface of these loaded gears and bearings, resulting in fractured gear teeth with a grainy, gray appearance, fractured output shaft and spalled bearings.

Replace the mainshaft overdrive gear and avoid lugging the engine.

Recommendation: Replace

Fig. 54

TRANSMISSION COUNTERSHAFT GEAR

The breakdown of tooth surfaces and the concentration of pitted and spalled areas, at the edges was caused by excessive driveline loading during extremely severe service operation. This is associated with the housing deflection which concentrates the contact at the edge of the teeth.

Recommendation: Replace

Fig. 55

BUMPS, BURRS AND SWELLS ON OPERATING TRANSMISSION GEAR TEETH

Careless handling of finished gears can cause bumps, burrs, and swells on the teeth.

Recommendation: Replace

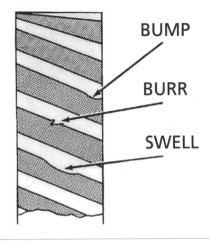

Fig. 56

APPEARANCE OF FAILURE

Poor tooth contact by a normal pattern shows up through a concentration of the polished surfaces at either end of the teeth. This results in a high-pitched whine at high shaft speeds.

Excessive deflection of the gear housing or gear misalignment probably caused the wear.

These wear patterns usually do not result in failure.

Recommendation: Check tooth contact and alignment. Reuse.

Fig. 57

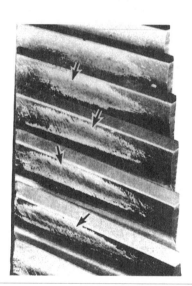

Fig. 58

Fig. 59

TEST YOURSELF

QUESTIONS

1. What are the three types of gear wear?

2. What causes pitting, spalling, and case crunching of the gear teeth?

3. What type of gear failure is caused by repeated, excessive loads that fracture the teeth normally at or near the tooth root?

4. What type of gear failure is identified by a gray and grainy fracture at or near the tooth root?

5. What causes rippling, ridging, and cold flow of the gear surfaces?

6. (True or false?) Incorrect backlash is a major cause of ring gear failures.

7. Discolored and distorted teeth on a ring gear is an indication of what?

8. (True or false?) Gear failure is rarely caused by operator abuse.

SHAFTS, AXLES, SPINDLES, AND UNIVERSAL JOINTS

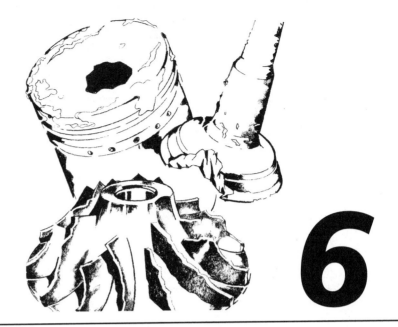

6

INTRODUCTION

This section on shafts, axles and spindles is divided into two broad categories:

- **General Category of Failures**
- **Specific Category of Failures**

GENERAL CATEGORY OF FAILURES

The general category applies to failures common to all shafts, axles and spindles regardless of the application and assembly conditions. First, the following failures are described:

- **Overload failures**
- **Bending fatigue failures**
- **Torsional fatigue failures**
- **Combined fatigue failures**
- **Impact failures**

The following specific causes of failure are then described:

- **Severe service**
- **Stress concentration**
- **Fretting and scoring**
- **Improper use**
- **Failure of other parts**

OVERLOAD FAILURE

High static loads tend to produce the type of failure shown. Although the shaft has not fractured, it has twisted and is out of alignment. Static load is the weight the shaft must support while sitting still. In this case, an overload probably caused the failure.

Recommendation: Replace

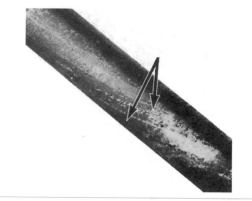

Fig. 1

FATIGUE FAILURE

A fatigue failure which is more common than static or impact, has a different appearance. The first phase of fatigue takes time. As a result, the progressive portion of the fatigue fracture (A) is smoothed by the continual rubbing between the two faces of the crack. That portion which failed instantaneously (B) however, has a rough surface.

Recommendation: Replace

Fig. 2

There are many variations of this basic pattern. This failure was caused by progressive fatigue cracking, originating at (A). The cracking progressed across most of the section before the final overload fractured the remaining metal. The probable cause of failure was repeated overloading.

Recommendation: Replace

Fig. 3

TORSIONAL FATIGUE

This failure was caused by a twisting load. Torsional (twisting) loads produce spiral types of failure. Notice the curved line from "A" to "B".

Recommendation: Replace

Fig. 4

Sometimes the evidence of fatigue is obscured by severe battering at the time of final fracture. This crankshaft failed initially by fatigue.

Recommendation: Replace

Fig. 5

SEVERE SERVICE

The deformation of these splines in the region of the fracture (arrows) would not occur if the fracture were caused by fatigue. The torsional, twisting during an overloaded operation caused the failure.

Recommendation: Replace

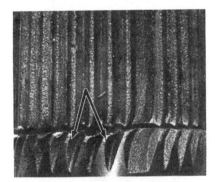

Fig. 6

FRETTING AND SCORING

This failure started in at arrow due to minute rubbing action between the shaft and another part of the machine.

Recommendation: Replace

Fig. 7

COMBINED FATIGUE FAILURES

Combined bending and torsional loads caused this tractor axle failure.

The outer surface of the axle was hardened. The crack began in the hardened surface at the top (A). The rough area in the center (B) indicates the area of instant fracture when the axle finally broke.

Recommendation: Replace

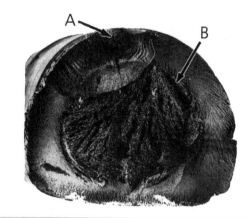

Fig. 8

IMPACT FAILURES

The face of a fracture caused by impact loading is usually gray, fibrous, and grainy, without evidence of progressive failure as in fatigue. The axle probably was struck by a single blow or overload causing immediate fracture.

Recommendation: Replace

Fig. 9

This transmission shaft shows the progressive stages of scoring. Scoring is caused by heat build-up between metal parts that rub together.

Recommendation: Replace

Fig. 10

IMPROPER USE

A number of irregularly shaped and spaced rack teeth were ground in beside the existing machined teeth. This was definitely performed in the field.

Recommendation: Replace

Fig. 11

SPECIFIC CATEGORY OF FAILURES

The specific category refers to specific applications. The following specific parts are described:

- **Crankshafts**
- **Oil pump shafts**
- **Transmission shafts**
- **Wheel spindles**
- **Universal joints**

CRANKSHAFTS

This failure is due to inadequate lubrication. The crank removed from an engine appears to be cracked. The journal surface was scratched and shows evidence of wear and exposure to high frictional temperature generated during operation. Large amounts of aluminum bearing material are welded to the surface of the rod journal.

Recommendation: Replace

Fig. 12

This main bearing journal of a crankshaft from a diesel engine was badly scratched by dirt particles which contaminated the lubricant.

Recommendation: Replace

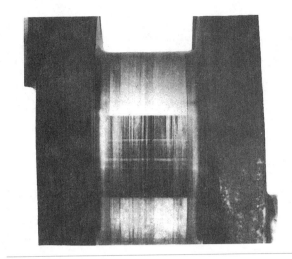

Fig. 13

OIL PUMP SHAFTS

This broken pump drive shaft was overloaded from drive belts that were too tight.

Recommendation: Replace

Fig. 14

This shaft of an oil pump drive gear shows fretting wear which has led to corrosion. Additional use could cause fatigue failure. Fretting is caused by a slight reciprocating movement between metal parts that are in close contact.

Recommendation: Replace

Fig. 15

TRANSMISSION SHAFTS

This gear and shaft show galling and overheating (discoloration on shaft beside galled area). Inadequate lubrication caused the failure.

Recommendation: Replace

Fig. 16

WHEEL SPINDLES

A large amount of wear is visible on both sides of this broken spindle.

Recommendation: Replace

Fig. 17

This fractured spindle, of a steering knuckle, was caused by fatigue cracks formed at the opposite sides of the spindle at the arrow.

Recommendation: Replace

Fig. 18

UNIVERSAL JOINTS

A universal joint is essentially a double-hinged joint which transmits torque at constantly changing relative angles. High loads, lack of lubrication and abrasive material are responsible for most of the damage exhibited by the universal joints.

Some recognizable signs of universal joint deterioration are:

- **Vibrations**

- **Universal joint looseness**

- **Universal joint discoloration due to excessive heat build-up**

A significant majority of universal joint failures are related to lubricating film breakdown. This may be caused by:

- **A lack of lubricant**

- **Inadequate lubricant quality for application**

- **Inadequate initial lubrication**

- **Failure to lubricate properly or often enough**

Failures which are not a result of lubrication film breakdown are associated with the following:

- **Installation**

- **Excessive universal joint angles**

- **Excessive operating speeds**

- **Overload**

Some common universal joint failures are described on the next four pages.

LACK OF LUBRICATION

This universal joint failed due to a lack of lubrication.

Recommendation: Replace

Fig. 19

END GALLING

This surface damage is galling or smearing. It occurs when surfaces rub against each other, get hot, and stick together. Small pieces then tear away from that surface and are welded to another surface area.

End galling results from inadequate lubrication, inadequate clearances, and operating at sharp angles and high speeds.

Recommendation: Replace

Fig. 20

BRINELLING

Brinelling of the universal joint is caused by another piece of metal being pressed into the damaged area shown. The fit of the parts was too tight in this example.

Recommendation: Replace

Fig. 21

SLIP SPLINE GALLING

Sliding splines require a uniform and complete initial lubrication around and along the length of the spline.

Spline surfaces can run dry if not properly lubricated. This causes localized galling (surface damage of rubbing parts being welded together from friction and burning that leads to early failure).

Recommendation: Replace

Fig. 22

FATIGUE

Three stages of journal fatigue failure are shown — early stages at left to severe spalling at right.

Fatigue failures result from applying too high a load for a given joint size and operating at too great an angle. Inadequate lubrication also contributes to early fatigue failure.

Recommendation: Replace

Fig. 23

Fatigue failures also show up on the inside diameter of a u-joint bearing cup.

Recommendation: Replace

Fig. 24

BROKEN JOURNAL

Journal fracture normally occurs at the base of the trunnion — the point of highest bending stress.

Excessive torques for given u-joint sizes and excessive shock loads cause this type of failure. Sudden drive shaft lock-up (as occurs during some equipment failure) causing excessive u-joint loads, high operating angles causing vibrations, and high torque fluctuations also lead to broken journals.

Recommendation: Replace

Fig. 25

TEST YOURSELF

QUESTIONS

1. What is the "static" load of a shaft?

2. (True or false?) A fatigue fracture has a smooth surface where the faces of the initial crack rubbed against each other and a rough surface at the point of final separation.

3. What is the probable cause of fatigue fractures?

4. (True or false?) The face of an impact fracture does not show the progressive failure of a fatigue fracture.

5. What is the usual cause of crankshaft failure?

6. What causes the majority of universal joint failures?

HYDROSTATIC TRANSMISSIONS

INTRODUCTION

This chapter deals with failures involving the following hydrostatic transmission components:

- **Thrust Plates or Fixed Swashplate**
- **Piston/Slipper Assembly**
- **Slipper Retainer**
- **Ball Guides**
- **Cylinder Block**
- **Bearing Plates**
- **Bi-Metal Bearing Plates**
- **Valve Plates**
- **Shaft Seal**
- **Charge Pump Assembly**
- **Displacement Control Valve**
- **Servo Sleeve and Piston**
- **Shafts**

Most failures shown are for heavy-duty hydrostatic systems. Failure appearances in light-duty systems are similar to those of heavy-duty systems. See the last section of this chapter for specific reference to failures in light-duty hydrostatic systems.

HEAVY DUTY HYDROSTATIC SYSTEMS

IMPORTANT: Hydrostatic systems require proper quantities and types of lubricant. These systems are very sensitive and parts fail if not properly lubricated.

THRUST PLATES OR FIXED SWASHPLATES

SMEARING OR GALLING

This condition is usually caused by a lack of lubrication; insufficient or improper fluid being used. Bronze material embedded into the thrust plate (metal transfer) indicates smearing or galling.

Recommendation: Replace

Fig. 1

SCORING

This plate was scored by abrasive contaminants suspended in the hydraulic fluid.

Scoring is indicated by fine scratches or grooves in thrust plate.

When these scratches can be detected by feel, fingernail or lead pencil, the plate should be replaced.

Recommendation: Replace

Fig. 2

PISTON/SLIPPER ASSEMBLY

SCORING

Fine scratches across the slipper face are caused by abrasive contaminants suspended in the hydraulic fluid. There is also some discoloration which usually indicates improper fluid, or large quantities of water in the fluid.

If scratches can be detected by feel with fingernail or lead pencil, the parts should be replaced.

Recommendation: Replace

Fig. 3

DAMAGED SLIPPER

Damage across the balance land usually begins as a deep scratch and erodes larger as high pressure escapes across this area. The original scratch is caused by a large particle of contamination.

Recommendation: Replace

Fig. 4

ROLLED

Overspeeding is indicated by the outer edge of the piston slipper being rolled over and in some cases the sides of the slippers being skinned up. This is usually caused by the unit overspeeding which causes the slipper to ride on its outer edge rather than across its entire face.

Recommendation: Replace

Fig. 5

CONTAMINATED SLIPPER

A solid particle of contamination embedded in the slipper face will cause scoring of the thrust plate.

Recommendation: Replace

Fig. 6

SMEARING OR GALLING

The damage to the entire face of this slipper was caused by lack of lubrication. There was insufficient or improper fluid in the system.

Recommendation: Replace

Fig. 7

DISCOLORATION

The discoloration on this piston outside diameter indicates the unit was subject to excessive temperature.

Recommendation: Replace

Fig. 8

PISTON FILLER SEPARATION

Separation is usually caused by the system being exposed, for long periods of time, to cavitation or high temperature. Cavitation occurs when excessive amounts of air or foam are present in the fluids and the air replaces the hydraulic fluid films. The air does not provide adequate support for the running surfaces. The piston filler may separate in smaller pieces or in one piece as shown. The filler insert is not replaceable.

Recommendation: Replace piston

Fig. 9

SLIPPER SEPARATION

A stuck piston can cause a slipper to separate from the piston. Excessive overspeed, contamination or lack of lubrication can cause this.

Recommendation: Replace piston

Fig. 10

SLIPPER RETAINER

DISCOLORATION

Discoloration on this slipper retainer indicates the retainer was subject to excessive temperature.

Excessive temperature may cause the retainer to distort or fracture.

Recommendation: Replace

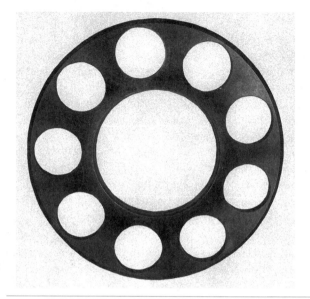

Fig. 11

SCORING

A severe wearing pattern (scoring) where the piston slipper contacts the retainer is shown.

This usually indicates the system was exposed to abrasive contaminants suspended in the hydraulic fluid. This same scoring may also be found on the inside diameter of the retainer where contacting the ball guide.

If this scoring can be detected by feel with fingernail or lead pencil, the part should be replaced.

Recommendation: Replace

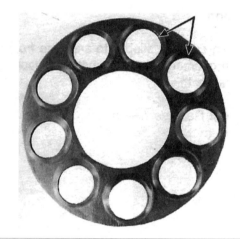

Fig. 12

LOW SYSTEM CHARGE PRESSURE

This slipper retainer was pulled down over the ball guide because the pressure difference between the unit case and the system charge pressure was too low.

Possible causes are high case pressure (restricted outlet), a plugged charge pump inlet filter, low reservoir level, or low speed-high pressure operation of the pump. In low speed-high pressure operation the charge pump cannot keep up with the flow demand from the pump, and charge pressure drops.

Recommendation: Replace

Fig. 13

BALL GUIDES

SCORING

Scoring indicates abrasive contaminants between the two mating seal parts. This contamination may have been introduced from outside the unit or could have been suspended in the hydraulic fluid.

When these scratches can be detected by feel, fingernail, or lead pencil the ball guide should be replaced.

Recommendation: Replace

Fig. 14

WORN

Worn areas around the lubrication holes usually indicate a lack of lubrication or abrasive contaminants in the hydraulic fluid.

Recommendation: Replace

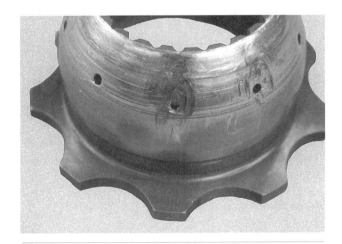

Fig. 15

CYLINDER BLOCK

PULLED BUSHING

This pulled bushing may have been caused by a seizure of the piston in the cylinder bore as a result of contamination, over-speed, or lack of lubrication.

The cylinder bore bushings are not replaceable.

Recommendation: Replace cylinder block

Fig. 16

FRACTURE

Excessive system pressures or excessive misalignment could have caused this fracture.

Recommendation: Replace

Fig. 17

BEARING PLATES

SMEARED-OUTER EDGE

This bearing plate shows smearing on the outer support area. Smearing is a metal transfer caused by severe friction between rotating parts.

Recommendation: Replace

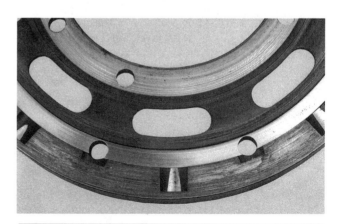

Fig. 18

EROSION

The cylinder block side of the bearing plate may show erosion on the balance land resulting in excessive internal leakage.

Recommendation: Replace

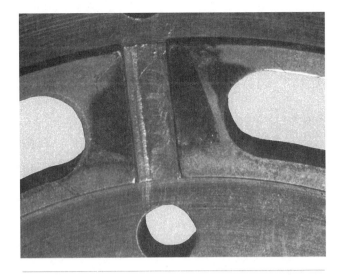

Fig. 19

GROOVED

Grooves between the kidney areas are usually caused by abrasive contamination in the high pressure circuit.

If the groove can be detected by feel with fingernail or lead pencil, the part should be replaced.

Recommendation: Replace

Fig. 20

TOTAL SMEARING

This bearing plate shows smearing across its entire surface, usually caused by abrasive contamination or by lack of lubrication.

Recommendation: Replace

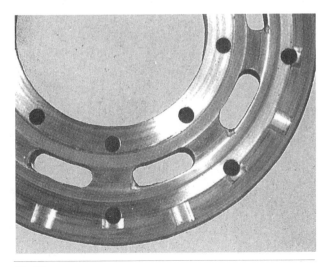

Fig. 21

DISCOLORATION AND SMEARING

This plate is starting to score because of lack of lubrication or improper fluid. The color appearance also indicates improper fluid or a chemical reaction to the bearing plate material.

Recommendation: Replace

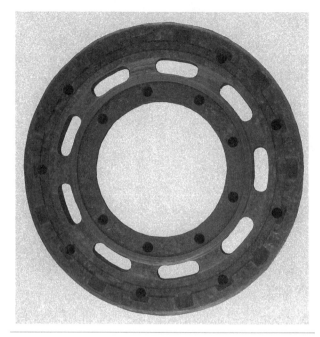

Fig. 22

DISCOLORATION

A bearing plate too dark in color (tarnished) usually indicates improper fluid or large quantities of water suspended in the fluid. Excess temperature may also cause discoloration.

Recommendation: Replace

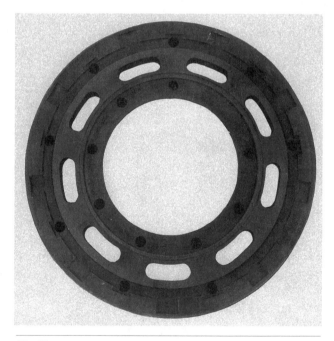

Fig. 23

The lighter area on the cylinder block side of this plate shows wear caused by plate movement against the cylinder block.

Excessive temperature and improper fluid may cause the plate to flutter causing a wear pattern on the cylinder block side of the plate.

Recommendation: Replace

Fig. 24

CAVITATION

Cavitation has eroded the kidney area of this bearing plate on the cylinder block side.

When cavitation occurs on the inner or outer balance lands the plate should be replaced.

Recommendation: Replace

Fig. 25

BI-METAL BEARING PLATES

Bi-metal bearing plates are made of bronze bonded to steel and have a different color. In general, the previously described guidelines also apply to the bi-metal plates.

TOTAL SMEARING

This bearing plate shows slight erosion due to cavitation in the kidney areas with smearing across the entire running surface. The probable cause is lack of lubrication.

Recommendation: Replace

Fig. 26

CAVITATION

This advanced stage of cavitation shows some bronze material eroded completely away from the steel base material of the plate at the kidney areas. This condition usually is caused by air bubbles trapped in the fluid.

Recommendation: Replace

Fig. 27

METAL SEPARATION

Bronze material removed from the bearing plate indicates extended cavitation in the system. High temperature may also contribute to metal separation from bi-metal plates.

Recommendation: Replace

Fig. 28

VALVE PLATE

CONTAMINATION

A particle of contamination embedded on the back side of the valve plate can cause the plate to lift and result in excessive internal leakage and damage to the bearing plate.

Recommendation: Replace

Fig. 29

SMEARING

Smearing is normally caused by a lack of lubrication, insufficient or improper fluid, or excessive temperature.

Recommendation: Replace

Fig. 30

GROOVING

Grooves and nicks between the kidneys and smearing on the bearing areas indicate abrasive contaminants are suspended in the hydraulic fluid.

Recommendation: Replace

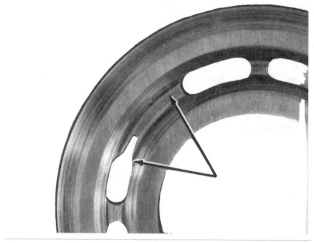

Fig. 31

SHAFT SEAL

GROOVING

The severe groove in the bronze rotating part of this seal indicates the seal was subject to abrasive contamination.

This type of failure is caused by excessive outside pressure against the mating seal parts. A non-vented gear box could cause the excessive pressure.

Recommendation: Replace

Fig. 32

SCORING

This scoring indicates abrasive contaminants between the two mating seal parts. This contamination may have been introduced from outside the unit or could have been suspended in the hydraulic fluid.

Recommendation: Replace

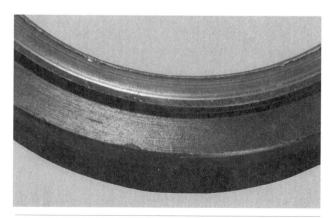

Fig. 33

FRACTURE

The fracture on the steel stationary part of this seal indicates the seal assembly was improperly installed.

Recommendation: Replace

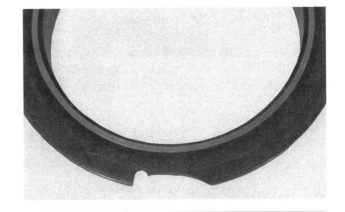

Fig. 34

CHARGE PUMP ASSEMBLY

FRACTURED INPUT FLANGE

This fracture indicates the fitting was tightened too much. A severe blow to the adapter may also cause this type of failure.

Recommendation: Replace

Fig. 35

DISPLACEMENT CONTROL VALVE

BROKEN SPRING

The spring pin that attaches the control shaft to the two pivots may fracture and result in a broken spring.

Recommendation: Replace

Fig. 36

BROKEN CONTROL SHAFT

When the external portion of the control shaft is broken it is a result of either too much torque on the nut which retains the control handle or improper handling during shipping or operation.

IMPORTANT: All displacement controls are factory adjusted, locked and sealed with a cap noted at top of valve shown. Do not disturb this adjustment.

Recommendation: Replace

Fig. 37

SERVO SLEEVE AND PISTON

FRACTURE

A fracture in the recess area usually indicates the sleeve was misused, such as dropping the unit on the servo sleeve.

Recommendation: Replace

Fig. 38

SCORING

Contamination usually causes scoring on the inside bore of the servo sleeve.

Recommendation: Replace

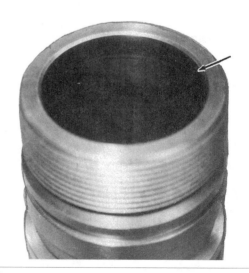

Fig. 39

Scoring in the servo piston is usually caused by abrasive contamination. If scratches can be detected by feel with fingernail or lead pencil, the part should be replaced.

Recommendation: Replace

Fig. 40

SHAFTS

WORN SPLINE

A worn spline usually results from an improper fit or misalignment of the mating coupling.

Recommendation: Replace

Fig. 41

FRACTURED INPUT TAPER

A broken taper is usually the result of an improper fit of the coupling to the shaft, or improper torque of the coupling retaining nut.

Recommendation: Replace

Fig. 42

WORN SPLINE

A worn spline normally is caused by excessive misalignment or torsional stress to the input shaft. Excessive load conditions may also cause this type of failure.

Recommendation: Replace

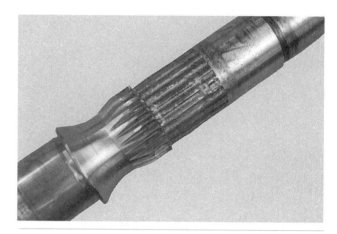

Fig. 43

LIGHT DUTY HYDROSTATIC SYSTEMS

LOW INPUT SPEED OPERATION

These slipper thrust faces have rounded edges due to rolling. This can be caused by charge pump failure, or low speed operation. Restricted reservoir, line, or filter; low oil in reservoir; or internal leakage from wear or misassembly could also cause the failure. A broken retainer and stuck pistons, result from continued operation with low system charge pressure.

Fig. 44

STUCK PISTON

Stuck pistons result from contamination in the system or damaged cylinder bore or piston surface. The slipper retainer is bent in the area of the stuck piston.

Fig. 45

VALVE PLATE/CHARGE PUMP HOUSING

The scoring of this valve plate and charge pump housing was probably caused by contamination in the system prior to start up or the filter not being maintained properly.

Recommendation: Replace

Fig. 46 — Valve Plate

Fig. 47 — Charge Pump Housing

TEST YOURSELF

QUESTIONS

1. What is scoring?

2. What causes scoring?

3. When should scored parts be replaced?

4. What is smearing or galling?

5. What causes smearing or galling?

6. What is cavitation wear?

7. What causes cavitation wear?

8. What causes the discoloration (or tarnishing) of plates?

9. What causes nicks on hydraulic components?

10. What causes the majority of failures in hydrostatic transmissions?

ANTI-FRICTION BEARINGS

INTRODUCTION

Failures of anti-friction bearings are due to a variety of causes, most linked to the following:

- **Contamination**
- **Improper lubrication**
- **Improper installation**
- **Careless handling**
- **Distortion and misalignment**
- **Severe service**
- **Vibration**
- **Electric current**
- **Defects in bearing material**

CONTAMINATION

Contamination is any foreign matter that will damage the bearing. Moisture and any type of abrasive, such as dirt or sand, will cause premature failure.

Abrasive contaminants and moisture rusted, scratched, and scored the bearing races. This type of damage may be prevented by using the correct lubricant, keeping the bearing clean while handling it and using new or undamaged seals.

Recommendation: Replace

Fig. 1

This bearing shows the effect of coarse abrasives on the raceways. Although it is difficult to picture, the dull, gray discoloration on the raceway surfaces contrasts with the bright finish of a new bearing.

Recommendation: Replace

Fig. 2

Foreign material caused excessive wear to these bearings. The roller ends are worn down to the indent and the ribs are badly worn.

The second type of damage caused by foreign material is pitting. Pitting is a type of fatigue failure and appears when small particles of the bearing separates from the surface. As surfaces of mating parts come in contact, repeated stress on these surfaces can cause pitting.

Recommendation: Replace

Fig. 3

Metal chips or large particles of dirt remaining in improperly cleaned housings are the most usual causes of trouble.

Fairly large particles of metal or dirt bruised and pitted the bearing race. Some of the indentations are so deep that the hardened surface of the bearing has fractured. The surface of the races would soon flake or spall (advanced pitting) with further operation of the bearing.

Recommendation: Replace

Fig. 4

Rust or corrosion is a serious problem in antifriction bearings. The high degree of finish on races and rollers makes them quite susceptible to corrosion damage from water.

Corrosion is often caused by condensation of moisture in the bearing housing because of temperature changes. Moisture or water may get in through damaged, or worn seals. Considerable damage also results from improperly washing and drying bearings when they are removed for inspection.

Recommendation: Replace

Fig. 5

This bearing shows more advanced damage from corrosion. The cones and the cups show spalling of the races at the areas where the heavy corrosion took place.

Recommendation: Replace

Fig. 6

Advanced corrosion damage first caused these pits and spalls. Then as the metal continued to break down, as rollers hit the edges of the spalled areas, the complete raceway surfaces in the loaded area were broken down or spalled.

Recommendation: Replace

Fig. 7

Water enters during unfavorable storage, through worn seals, or by leakage through gaskets or covers on housings. Results are heavy corrosion, or rusting.

Recommendation: Replace

Fig. 8

IMPROPER LUBRICATION

Proper lubricant is important for a satisfactory bearing operation. It can be a complete lack of lubricant or too small a supply that causes bearings to fail. It may be the wrong kind of lubricant, or wrong grade or weight. Always use the lubricant specified by the manufacturer.

Too much lubricant that was too solid was used in this bearing. The rolling elements slide across the surface, rather than roll, smearing the metal. The thick lubricant slowed the rolling elements enough to start the sliding. As the metal surfaces became smeared, the rate of wear increased. This damage can also result from inadequate lubricant.

Recommendation: Replace

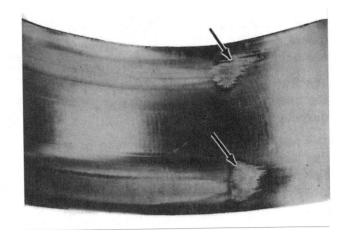

Fig. 9

Another form of surface damage is progressively shown in A, B, C and D. The first visible indication of trouble is usually a fine roughening of the surface. Later spalling will begin from fine cracks.

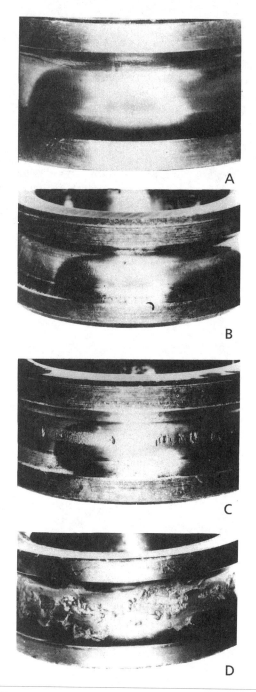

A

B

C

D

Fig. 10

This bearing shows discoloration and softening due to excessive heat.

Recommendation: Replace

Fig. 11

A bearing may glaze (top) or balls may become welded to the ring (bottom).

Recommendation: Replace

Fig. 12

Fig. 13

This broken cage was caused by intermittent lack of lubricant. At times, because of certain operating conditions, the oil fails to get to the bearing for a period long enough to cause light scoring of rollers and rib. This causes pressure on sides of cage pockets. After this has repeated a number of times the cage finally starts to crack and then break up as shown.

Recommendation: Replace

Fig. 14

IMPROPER INSTALLATION

Improper installation will also cause premature bearing failure. Excessive preload or tight adjustment can cause damage similar to inadequate lubrication damage. Frequently the two causes may be mixed so that a careful check is required to determine the real trouble.

There are three types of damage resulting from improper installation. The first is a split race caused by forcing the bearing onto a shaft too large for the inside diameter of the bearing race.

Recommendation: Replace

Fig. 15

This outer race was damaged by "fit rust" or fretting corrosion. This results when the outer race fits too loosely in its housing. Sanding or removing this rust will only increase the loose fit. The bearing must be replaced.

Recommendation: Replace

Fig. 16

Creep wear is caused by too loose of a fit between the shaft and the inside bore of the bearing. As this wear progresses, the inner race turns faster creating more friction and heat and leads to eventual bearing failure.

Recommendation: Replace

Fig. 17

This bearing race is badly spalled from fatigue of metal. This fatigue occurred prematurely and could be eliminated by removing the preload or reducing it to the proper amount.

Recommendation: Replace

Fig. 18

CARELESS HANDLING

Bearing damage can be caused by careless handling, by improper service technique, and by use of improper tools during installation.

These bearings received improper service. Example A shows nicks in an outer race caused by using a hammering tool (drift) to drive the bearing. Example B is of cracks caused by striking the race with a hammer.

Recommendation: Replace

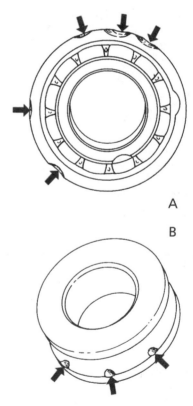

A

B

Fig. 19

Hammering also damages the inner race as shown in Example C. When the race is struck a severe blow, the force is transmitted through the rolling elements to the other race and chips it. Another result of using improper tools is a damaged seal on the sealed bearing in Example D. A drift slipped and gouged this seal. The seal's effectiveness is reduced and the separator is probably put in a bind.

Recommendation: Replace

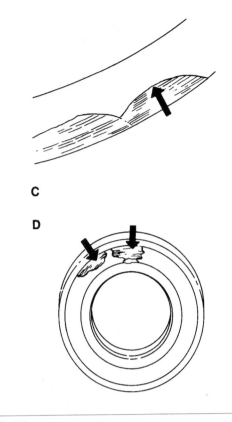

C

D

Fig. 20

This bearing was dropped. It landed so that the cage was bent at the large end. This cage distortion causes the roller to bind in the cage and distort.

Recommendation: Replace

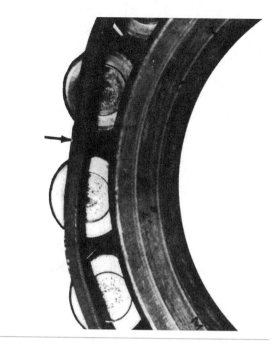

Fig. 21

The cage was damaged during installation because proper tools were not used. It is apparent that a bar or drift was used against the cage instead of the cone face to drive the cone on the shaft.

Recommendation: Replace

Fig. 22

This fatigue damage was caused by an impact during either handling or mounting. An impact leaves a depression that can initiate premature fatigue.

Recommendation: Replace

Fig. 23

Fig. 24

The nicks in these cups were caused by the cone being in a cocked position with respect to the cup. The ends of some of the rollers dug into the cup surfaces. The roller edges were flattened and metal pushed up on the body of the rollers at these points. Because of the cocking action or tipping, the rollers were forced down on the edges of the cone race causing impressions at the extreme large and small ends of the cone raceway.

Recommendation: Replace

Fig. 25

The bar or tool used to drive in this cup slipped and dug into the surface.

Recommendation: Replace

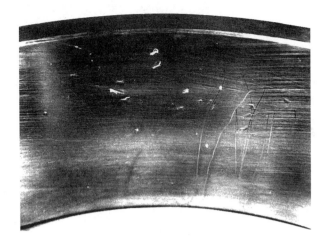

Fig. 26

Tool marks extending out to the edge of the cup seat causing a high spot resulting in spall or fatigue at this point in the cup.

Recommendation: Replace

Fig. 27

These inside diameter and outside diameter markings in a cup were caused by a high spot in the housing. The inside diameter is spalled at the area and the outside diameter shows heavy contact at the corresponding spot.

Recommendation: Replace

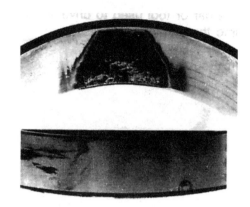

Fig. 28

MISALIGNMENT

Misalignment is usually caused by a bent shaft, or foreign matter between the bearing and its seat. Notice the paths worn in the races and the worn shape of the balls. Misalignment in roller or needle bearings usually results in extreme pressure on the races and rollers, and premature fatigue failure. The cause of failure must be determined and corrected or the same damage will result when new bearings are installed.

Recommendation: Replace

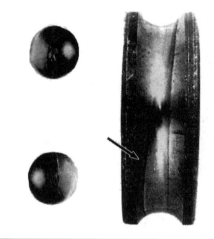

Fig. 29

Excessive end play in a bearing caused this condition.

Recommendation: Replace

Fig. 30

SEVERE SERVICE

Extremely heavy impact loads of short durations can result in impressions in the bearing races and sometimes even fracturing the races and rollers.

If thrust bearings are set with end play, there is a pounding action set up as the wheel goes over uneven surfaces. These rapid short impacts pound the rollers into the races and ultimately will fracture the race.

Recommendation: Replace

Fig. 31

The first indications of fatigue failure may be noisy running and increased vibration. These bearing races show the metal surfaces of the races have flaked away. This flaking is caused by the effects of excessive speed and load.

Recommendation: Replace

Fig. 32

VIBRATION

Most anti-friction bearings are rolling while under load. This bearing, however, was stationary while subjected to vibration. The depressions result from a combination of wear and impacts caused by this vibration. When this bearing is subjected to a rolling load, it will fail rapidly.

Recommendation: Replace

Fig. 33

This wear is caused by the rollers sliding back and forth on the race while the bearing or race is stationary. A groove is worn into the race by this sliding of the roller back and forth across the race. Vibration causes the sliding movement. The vibration present may cause enough movement to produce some of this wear.

The heavier markings and particularly the deeper and sharper narrow grooves will cause noise and roughness in the bearing.

Recommendation: Replace

Fig. 34

ELECTRIC CURRENT

Where bearings are used in the presence of electricity, potential damage can occur if the electrical current passes through the bearing.

Several electrical arcs that occur as the bearing turns melt the surface metal and lead to bearing failure. This pattern normally repeats itself many times around the diameter of the ring and race.

Recommendation: Replace

Fig. 35

Electric pits occur each time the current breaks in its passage between raceway and roller. This shows a series of electric pits. Amperage rather than voltage governs the amount of damage and both alternating and direct current cause damage.

The cause of the electrical leak must be found and corrected or it will occur again with a new bearing.

Recommendation: Replace.

Fig. 36

TEST YOURSELF

QUESTIONS

1. What causes contamination in anti-friction bearings?

2. What is probably the best way to prevent contamination?

3. (True or false?) Worn seals are seldom a cause of bearing failure?

4. (True or false?) Forcing a bearing onto a shaft larger than the inside diameter of the bearing can split the bearing race.

5. (True or false?) Too loose of a fit between the shaft and the inside diameter of the bearing is not a serious problem.

6. (True or false?) Many bearing failures are caused by the bearings being damaged during installation.

7. What causes misalignment failures?

8. (True or false?) Excessive preload can cause fatigue and spalling in a bearing.

BELTS AND CHAINS

9

INTRODUCTION

Belts and chains are flexible members for transmitting power. They have these advantages over other power transmitting devices:

- **They are suitable for relatively large center distances**
- **Belts absorb vibration and shock**
- **Belts may be quiet**
- **They have a long life, if properly maintained (usually not longer than other devices)**

V-BELTS

Belts can fail by wearing out, as a result of damage, or they can fail prematurely as a result of defective material in the belt.

V-belt failures are a result of the following causes:

- **Cracking**
- **Rupture**
- **Tear**
- **Burn**
- **Gouges**
- **Wear**
- **Internal Cord Failure**
- **Cuts**
- **Peeling, Fraying, Chewing**

CRACKING

Excessive cross-cracking on the base of a belt having little or no side wear indicates a defective belt only if it has run a very short time. This is a normal failure for belts that have run a long time at light load.

Recommendation: Replace

Fig. 1

This cracking is frequently caused by belt slippage causing heat build-up and gradual hardening, or by the belt running on sheaves that are too small. This is also a normal failure for belts that have run a long time.

Recommendation: Replace

Fig. 2

RUPTURE

The fabric rupture of this belt could have been caused by operating the belt over badly worn sheaves, too much tension forcing the belt down into the grooves, or objects falling onto the sheave groove while the drive is operating.

Recommendation: Replace

Fig. 3

This illustration shows a belt pulled apart. The cause of it could have been an extreme shock load, the drive being under extreme shock load or the belt coming off the drive.

Recommendation: Replace

Fig. 4

Ruptured cords are shown. Possible causes are foreign material damage or too much tension.

Recommendation: Replace

Fig. 5

TEAR

The cover tear of this belt is an example of damage caused by the belt accidently coming into contact with some part of the machine. In many cases, such failure is due to belts running too loose, allowing them to throw out centrifugally and rub on parts of the machine.

Recommendation: Replace

Fig. 6

BURN

The burned sides and bottom of this belt were caused by the belt slipping under a starting or stalling load.

Recommendation: Replace

Fig. 7

This belt shows the results of a spin burn. The driven pulley stalled from an overload or improper belt tension burning the belt when the drive sheave continued to run.

Recommendation: Replace

Fig. 8

GOUGES

Except for the gouged edge, the entire circumference of the belt shown is in new condition. Damage was caused by either a damaged sheave or interference with some part of the machine.

Recommendation: Replace

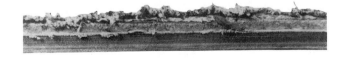

Fig. 9

The failure was caused by a gouge. Shock load at time of damage broke or weakened the belt causing the failure.

Recommendation: Replace

Fig. 10

This belt shows a gouge mark on the inside center of belt and the wrapper has started to peel. This indicates that some foreign object came in contact with the belt while in operation, causing the damage.

Recommendation: Replace

Fig. 11

WEAR

This badly worn belt is the result of long operation without enough tension. The sides are worn and slightly burned completely around the belt.

Fig. 12

Recommendation: Replace

Severe wear on the corners and surfaces of this belt shows that the belt rubbed against an obstruction.

Recommendation: Replace

Fig. 13

Worn belt sides, due to constant slip, are shown. The belt was probably improperly tensioned.

Recommendation: Replace

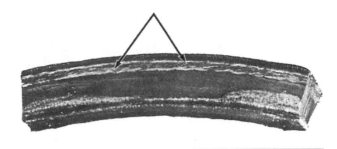

Fig. 14

The extreme wear pattern on the top corner and bottom corner of this belt indicates the belt was rubbing on some obstruction.

Recommendation: Replace

Fig. 15

This wear was due to the misalignment of the drive. Notice how both plies have been completely worn off of one sidewall while the other side of the belt shows normal wear.

Recommendation: Replace

Fig. 16

INTERNAL CORD FAILURE

This belt was overstressed and the inner cords were broken, causing premature failure. This belt was damaged during installation when it was roped over the sheave flange without releasing the tension.

Recommendation: Replace

Fig. 17

CUTS

The bottom cut on this belt could have been caused by the belt running over the sheave and coming off, foreign material falling into the drive making the belt come off, or the belt being forced over the sheave flange during installation without slacking off the drive.

Recommendation: Replace

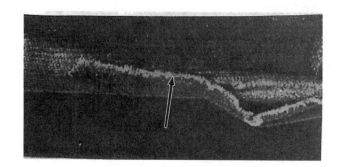

Fig. 18

The cut on the corner of this belt was caused by a sharp object, such as a belt guide coming into contact with the belt while in operation.

Recommendation: Replace

Fig. 19

This belt was damaged by the belt climbing out of a sheave groove while in operation. Notice the large section of belt cut out along the lower inside of the belt. The clean cut section indicates the belt was very tight when it climbed on the sheave.

Recommendation: Replace

Fig. 20

PEELING, FRAYING, CHEWING

The outer wrapper has started to peel on the inside of this belt. This type of damage could occur during installation or in service due to belt climbing over the sheave flange caused by improper adjustment or alignment.

Recommendation: Replace

Fig. 21

The inside cover of this belt has started to fray indicating the bottom of the belt was operating against some foreign object in the sheave grooves, such as mud, stalks or the sheave groove was wrapped with weeds.

Recommendation: Replace

Fig. 22

This belt shows that some foreign object entered the sheave, chewing up the cover or wrapper and causing the belt to turn over in sheave grooves. The lower portion of belt shows wear indicating the belt had been run in a turned over position in the sheave. This staggered pattern of cuts may have occurred when the belt turned over.

Recommendation: Replace

Fig. 23

BANDED V-BELTS

Banded V-belts were developed to solve troublesome problems on those drives where belts whip, turn over or come off drives.

Banded V-belts stay on the drive, eliminating downtime as well as the maintenance cost involved in reinstalling belts on the sheaves. These belts consist of multiple V-belts permanently vulcanized together with a tie band to minimize the possibility of its turning over or coming off the sheaves.

Banded V-belts, however, have some wear problems which are unique.

RIDING OUTSIDE OF SHEAVE GROOVE

Possible misalignment, lack of tension, or foreign object forced one strand of the belt from sheave groove.

Recommendation: Replace

Fig. 24

Riding outside of sheave grooves is a possible cause of the belts separating completely from the band.

Recommendation: Replace

Fig. 25

BOTTOMS OF BELTS CRACKING

The belts are cracked by running on sheaves that are too small, heat build-up caused by slippage and gradual hardening of the undercord, or have run a long time.

Recommendation: Replace

Fig. 26

TOP OF BAND FRAYED, OR DAMAGED

This damage was caused by an obstruction on the machine interfering with normal belt operation.

Recommendation: Replace

Fig. 27

BAND SEPARATING FROM BELTS

Worn sheaves probably caused this damage.

Recommendation: Replace

Fig. 28

HOLES OR BLISTERS IN TIE BAND

Accumulation of trash and foreign material between belts caused the damage.

Recommendation: Replace

Fig. 29

SPECIAL V-BELTS

The following illustrations refer to the failure of V-belts in the snowmobile power train.

FRAYED EDGES

If drive belt wears rapidly exposing frayed edge, cord is mis-aligned. Engine mounting bolts could also be loose allowing engine to twist and misalign belt.

Recommendation: Replace

Fig. 30

NARROW WEAR IN ONE SECTION

If drive belt is worn narrow in one section, excessive slippage is indicated due to a stuck or frozen track. Too high an engine idle speed could also be indicated if several narrow sections are evident.

Recommendation: Replace

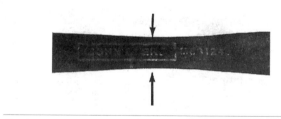

Fig. 31

Drive Belt Disintegration

Drive belt disintegration could be caused by a misaligned belt, using an incorrect belt or oil on sheave surfaces. If badly misaligned, drive belt will roll over at high speed causing belt disintegration.

Recommendation: Replace

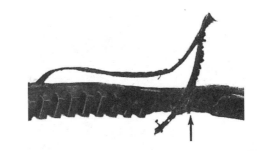

Fig. 32

DRIVE BELT WITH SHEARED COGS

A drive belt having sheared cogs could indicate violent engagement of drive sheave caused by binding or improperly installed drive sheave components. This is typical flex fatigue for this construction.

Recommendation: Replace

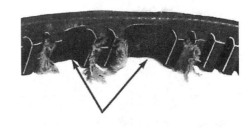

Fig. 33

FLAT BELTS

Common wear on flat belts is shown in the following illustrations:

FLAT BELT TEARS

An improperly aligned pulley causes a flat belt to climb the inner end of the engine drive pulley, tearing the edge of the belt. One of these tears may eventually cause the belt to tear completely through.

Recommendation: Replace

Fig. 34

FLAT BELT BURN

This shiny appearance and badly burned area were caused by operating the drive belt too loosely. Note the notch in the edge of the belt where it slid sideways against the flange of the drive pulley, while slipping and burning the center section.

Recommendation: Replace

Fig. 35

CHAINS

While this section deals primarily with roller chain, the reader should be aware that roller chain is just one of six basic types of precision chain. All six types are illustrated on the following page. Information regarding wear prevention in this section applies to all chains.

John Deere FOS *Belts and Chains* contains descriptions and applications for the other types of basic chains.

Standard roller chain has alternate roller links and pin links. Each roller link consists of two side bars, two bushings, and two rollers. Each pin link consists of two side bars and two pins.

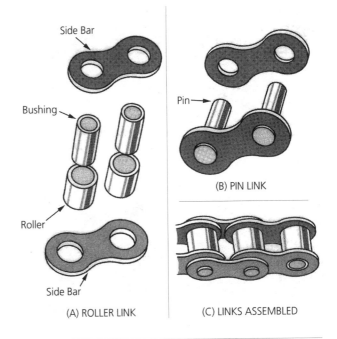

Fig. 36

Offset link roller chain uses links which are both a roller link and a pin link combined.

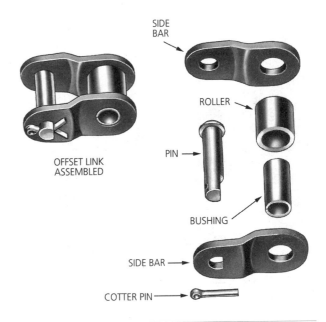

Fig. 37

ROLLER CHAIN

Fig. 38A

ROLLERLESS CHAIN

SILENT CHAIN

Fig. 38B

DETACHABLE LINK CHAIN

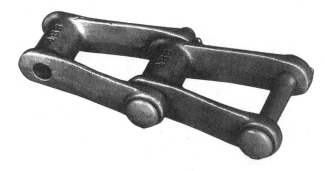

PINTLE CHAIN

Fig. 38C

BLOCK CHAIN

Fig. 38D

All chain will wear until it reaches its wear limitations and fails by breaking. Many factors contribute to this chain wear including:

- **Abrasives in the environment**
- **Overspeeding of the chain drive**
- **Overloading of the chain drive**
- **Misalignment of the drive components**

Probably the **two most major** causes of chain wear are:

- **Lack of lubrication**
- **Improper adjustment**

LACK OF LUBRICATION

Lack of lubrication can cause rusting and pitting which weakens the chain by increasing the wear on the pins and bushings.

Recommendation: If minimal pitting or rust is present, clean, lubricate and reuse. If severe pitting or rust is present, replace chain.

CADMIUM PLATED CHAIN (RUSTED)

STAINLESS STEEL CHAIN (RUST DISCOLORATION ONLY)

STAINLESS STEEL CHAIN (NO RUST)

STANDARD CHAIN (BADLY RUSTED)

Fig. 39

The wear on the pins of this link was caused by lack of lubrication.

Recommendation: Replace

Fig. 40

The wear on the rollers and bushings (bold arrows) of this roller link was caused by lack of lubrication.

Recommendation: Replace

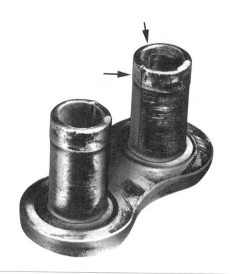

Fig. 41

Lack of lubrication can also cause galling as shown on these chain pins. Friction has caused particles of metal to be torn away as the chain rotated around the sprockets.

Recommendation: Replace

Fig. 42

IMPROPER ADJUSTMENT

Chains stretch as they are used. This is normal as material is being worn off the pin and bushing bearing surfaces. This wear occurs as the chain is flexed over the sprockets while under load. The slack caused by the stretch must be taken up to keep the chain from jumping off the sprockets.

See the manual of the machine the chain is used on for correct adjusting procedures for that particular chain drive.

A joint of a chain with the proper slack flexes only twice under load during each trip around the loop: 1) as it comes off the driven sprocket and 2) as it enters the drive sprocket

When there is no slack in the chain each link flexes each time it enters and leaves each sprocket. A chain too tight **accelerates** normal wear.

When there is too much slack the chain will whip and eventually jump off the sprockets.

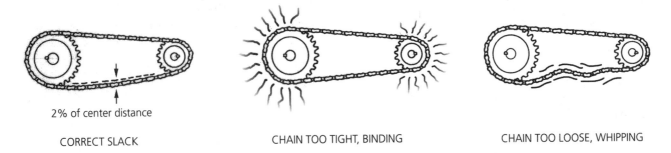

2% of center distance

CORRECT SLACK   CHAIN TOO TIGHT, BINDING   CHAIN TOO LOOSE, WHIPPING

Fig. 43

MEASURING CHAIN WEAR

Special gauges are available to check wear of some chains. The gauge illustrated shows the chain is worn out when both points enter pin recesses.

Wear on single-pitch roller chain can also be determined by measuring the old chain and comparing the measurement with that of a new chain. If the difference in measurements is 3% of the new chain measurement, the chain should be replaced.

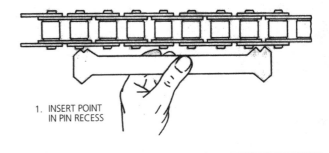

2. WHEN POINT ENTERS PIN RECESS HERE, CHAIN IS WORN OUT

1. INSERT POINT IN PIN RECESS

Fig. 44

TEST YOURSELF

QUESTIONS

1. (True or false?) Cracking is a normal failure for a belt that has been in service for a long time.

2. Give one reason for a belt rupturing. (Three were given in the text.)

3. What causes a cover tear in a belt?

4. What causes a belt to burn?

5. What causes cuts on the inside cover of a belt?

6. What causes fraying on the inside cover of a belt?

7. What causes "galling" or the tearing away of metal particles from a chain?

8. (True or false?) When properly adjusted there should not be any slack in the chain drive.

TRACKS AND TIRES

TRACKS

This section describes the failure of the following track under-carriage parts:

- **Links**
- **Pins and bushings**
- **Sprockets**
- **Rollers**

LINKS

If the link pin boss comes in contact with the roller flange, wear occurs on both the link and the roller flange. Severe wear may allow pins to become loose and fall out.

Loose hardware between the link and the shoes also wears links and can enlarge bolt holes.

Recommendation: Replace

Fig. 1

PINS AND BUSHINGS

A completely worn bushing results from not turning after one side wears. This shows both a bushing broken through and one not yet broken (right).

Recommendation: Replace

Fig. 2

SPROCKETS

Sprocket wear depends on the loading, terrain, abrasiveness and moisture content of the soil. Sprockets wear most uniformly and slowly if the track pitch and sprocket pitch are the same, as is the case with all new parts.

If the sprockets become packed with dirt or clay, the sprocket tooth tips wear due to a mismatch between the sprocket and track pitch. Also, with the increased track pitch caused by pin and bushing wear, the bushing climbs out over the sprocket tips causing wear.

Recommendation: Replace

Fig. 3

This sprocket shows excessive wear. The even wear across the sprocket indicates the sprocket may have been reversed. Chips in the teeth indicate the sprocket must have been hitting the rear roller or a stone lodged in the track frame.

Sprocket and other track parts wear more quickly through reverse and side hill operation and misalignment. When the bushings slide across the root of the sprocket teeth, or from side to side, it wears the sprocket root. If the bushings are loose, they will rotate while entering and leaving the sprocket, causing rotating wear.

Recommendation: Replace

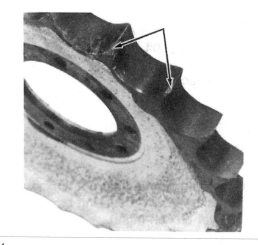

Fig. 4

ROLLERS

Rollers wear internally or externally. The internal roller components are permanently lubricated and sealed and should last the life of the roller, however, side hill operation or parking sideways on a slope puts pressure on the seals and can cause a loss of lubricant.

Heavy loading on the flange from the side hill operation causes the wear. Tread wear results as the links rub against the roller tread. This shows excessive tread wear and damage from an impact (large arrow). The top flange is also thinner, probably resulting from side hill operation or track misalignment (small arrows).

Recommendation: Replace

Fig. 5

The excessive wear of this carrier roller was caused by track misalignment, side hill operation or a bent shaft.

Fig. 6

Recommendation: Replace

Fig. 7

SPECIAL TRACKS

The following section refers specifically to snowmobile tracks:

CABLE AND EDGE DAMAGE

Each track has two woven steel cables running completely around the track. They may be ripped partially out or, if the entire edge is damaged, may be completely gone. Damage is most frequently caused by tipping the snowmobile on its side to clear the track or allowing the track to come in contact with an abrasive surface.

Recommendation: Replace

OBSTRUCTION AND IMPACT DAMAGE

Apparent cuts, slashes or gouges in the surface of the track are caused by obstructions such as broken glass, sharp rocks or buried steel. This commonly occurs during rapid acceleration or side skidding over foreign objects.

Fig. 8

Recommendation: Replace

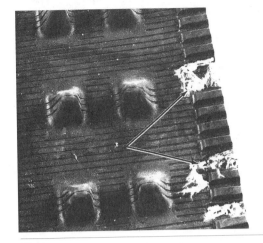

Impact damage to the edge of the track is usually caused by frequent riding on rough or frozen ground or on ice. Insufficient track tension, allowing the track to "pound" against the track stabilizers, may also cause impact damage. Excessive travel over buried logs may lead to damage.

Recommendation: Replace

Fig. 9

FACE DAMAGE

Excessively worn track face or grouser bars are caused by operating on extremely rough and dry terrain such as non-snow covered fields, railroad and highway right-of-ways, and gravel roads.

Fig. 10

Recommendation: Replace

LUG DAMAGE

Damage to the sides or rear edges of the lug is usually caused by lack of snow lubrication and excessive track tension.

Recommendation: Replace

Fig. 11

Ratcheting damage to the lug top results from insufficient track tension, pulling too great a load, or frequent prolonged periods of rapid acceleration.

Fig. 12

Recommendation: Replace

TRACK TENSION DAMAGE

Overtightening the track causes the three rear idler wheels to push excessively into the track. This will cause the fabric to break and the track to become "fuzzy" on the face from rear idler wheel pressure.

Recommendation: Replace

Fig. 13

Operating with a loose track allows the outer edge to flex. Some wear on the driving lugs may also be visible. Excessive weight without heavy-duty rear suspension can also cause the track to flex and break the edge.

Recommendation: Replace

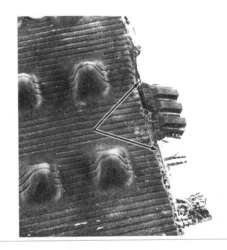

Fig. 14

TIRES

This section describes the following tire failures:

- **Fabric breaks**
- **Rubber checks**
- **Wear**
- **Tread wiping**
- **Tread and sidewall cuts**

FABRIC BREAKS

The small fabric break can be repaired.

Recommendation: Repair and reuse.

Fig. 15

Most fabric breaks are caused by hitting some object, which puts too much shock on the fabric.

Recommendation: Replace

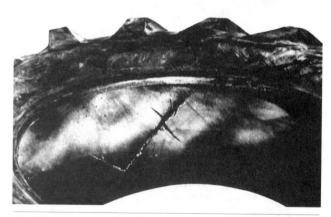

Fig. 16

The force is greater if the impact occurs at high speeds, or if the tires are overinflated. At high inflation pressures, damages may occur which later result in a large "X" or diagonal break which may extend from bead to bead.

Even at correct tire inflation, a severe localized blow, such as contact with a sharp rock or tree stump, can result in cord breakage like those shown.

Recommendation: Replace

Fig. 17

A special kind of tire break is caused by water freezing and expanding in the tire.

Recommendation: Replace

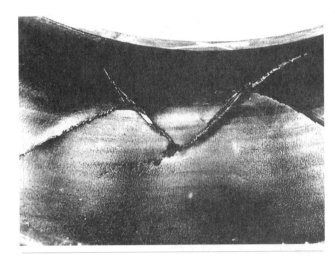

Fig. 18

On farm tractors, the furrow tire may be distorted by the tilt of the tractor. This puts a severe folding action on the tire, especially if the inflation is too low, causing the inner sidewall of the tire to finally separate in a series of breaks.

Recommendation: Replace

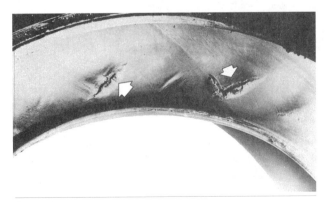

Fig. 19

This action may also cause a series of cracks at the edge of the tread bars, sometimes extending into the sidewall as shown.

Recommendation: Replace

Fig. 20

RUBBER CHECKS

Small checks or cracks on the rubber sidewall are usually caused by tires inflated to high pressures. Normally this is an appearance condition only and will not affect the service life of the tires.

Recommendation: Reuse

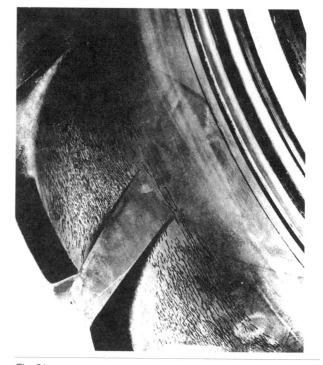

Fig. 21

WEAR

Tractor tires with too few wheel weights or too much inflation will wear the tread bars rough, or will snag and cut the bars, when subjected to severe service on abrasive surfaces. Sudden engagement of the clutch in starting also causes this type of tread wear. Tread bars are cut and worn on the leading or forward edge.

Recommendation: Reuse; if condition is severe, replace

Fig. 22

Operating a tractor tire over corn stubble causes wear and can puncture the tire. To avoid or limit stubble damage, adjust the row spacing of the rear tires so neither tire rides over the stubble. Avoid spinning the wheels where they do contact the stubble.

Recommendation: Replace

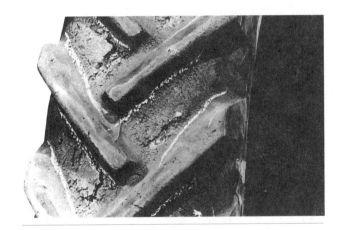

Fig. 23

TREAD WIPING

Tractor tires operating on hard roads with low inflation pressure cause an undesirable distortion of the tire. The tread bars "squirm" while going under and coming out from under the load. On abrasive or hard surfaces, this action "wipes" off the rubber of the tread bars or lugs and wears them down prematurely and irregularly.

Recommendation: Inflate properly and reuse

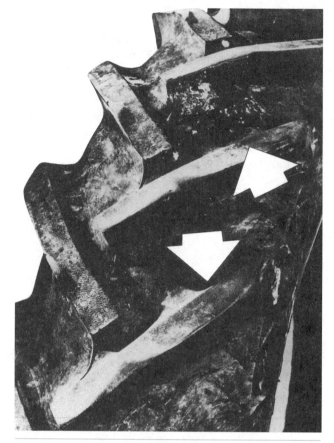

Fig. 24

TREAD AND SIDEWALL CUTS

Many tires injured by cuts or snags can be repaired and continued in service. Cuts or breaks that enter into or expose the cords in the body of the tire should be repaired promptly.

Recommendation: Repair and reuse.

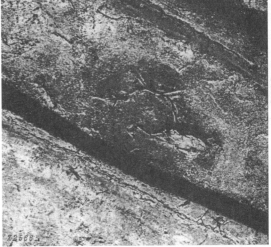

Fig. 25

If not repaired, moisture and foreign material can enter the injury and deteriorate the cords.

Recommendation: Replace

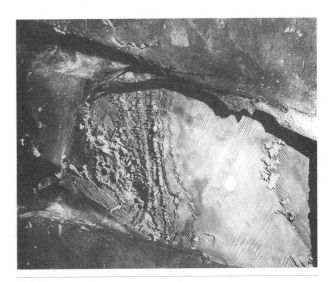

Fig. 26

TEST YOURSELF

QUESTIONS

1. What is a common cause of sprocket tooth tip wear?

2. (True or false?) Side hill operation accelerates undercarriage wear.

3. (True or false?) Operating a snowmobile on non-snow surfaces seldom damages the track.

4. (True or false?) Too much or too little tension can damage both the fabric and the lugs of a snowmobile track.

5. What causes most fabric breaks in tires?

6. What causes small checks or cracks on the sidewalls of tires?

7. (True or false?) Driving a tractor with under-inflated tires on a hard road will wear the tread bars or lugs prematurely.

8. (True or false?) Tires with cuts or snags that enter into or expose the cords in the body should always be replaced rather than repaired.

MISCELLANEOUS FAILURES

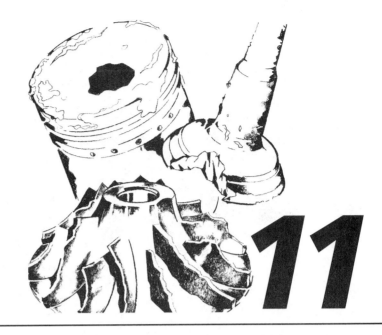

11

INTRODUCTION

This section deals with miscellaneous failures of the following parts:

- **Pump rotors**
- **Pump impellers**
- **Disk blades**
- **Planter finger pickup**
- **Spark Plugs**
- **O-rings**
- **Intercooler brace**
- **Fasteners**

PUMP ROTORS

One of the most common causes of rotor damage is abrasive wear. This shows various forms of damage to a hydraulic pump rotor. Vanes are gray and rotor slot wear is more than 0.002 inch (0.05 mm). Vane tips are worn and the rotor seized. The cam ring has two abrupt chop marks 180° apart. One vane broke from the impact of striking the steps in the cam ring.

Recommendation: Replace

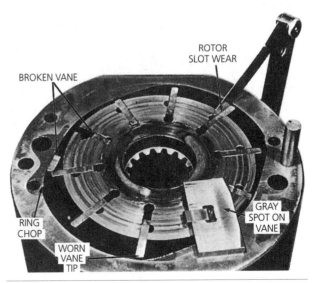

Fig. 1

A worn cam ring caused rotor and shaft spline wear (arrows). Vanes bouncing on a cam-ring surface badly worn from abrasives causes spline wear.

Recommendation: Replace

Fig. 2

Contaminants in the oil wore the vane tips (bottom). In normal operation the vane is held against the ring and a film of oil lubricates the vane tips and cam ring.

Recommendation: Replace

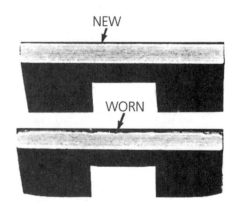

Fig. 3

Contaminated fluid can damage a pump in many ways. Solid particles of dirt and sand in the fluid act as an abrasive on the closely fitted parts. This causes abnormal wear on the parts.

Sludge is formed by the chemical reaction of fluid to excessive temperature change or condensation. It builds up on the pump's internal parts and eventually plugs the pump. If the pump is plugged on the inlet side, it will be starved for fluid and the heat and friction will cause the pump parts to seize.

Recommendation: Replace

Fig. 4

PUMP IMPELLERS

Abrasive wear occurs when hard particles slide or roll under pressure across a surface, or when a hard surface rubs across another surface. Surface projections in the harder object scratch or gouge the softer material.

A new impeller (top) can be compared to the worn impeller (bottom) which comes from an engine with sand in the cooling system. Erosive action of the abrasive particles moving at high speed progressively rounded the vane corners, gradually changing its shape.

Recommendation: Replace

Fig. 5

Fig. 6

Another form of impeller damage is cavitation corrosion. Cavitation corrosion or erosion is caused by a relative motion between a metal surface and a liquid. Small cavities of air in the liquid form and then collapse. This produces high impact forces on the metal, resulting in cavitation pits.

The water pump impeller (top) was damaged by cavitation. Another case of cavitation attack on pump impeller is shown (bottom).

Recommendation: Replace

Fig. 7

Fig. 8

This bronze pump impeller bushing shows corrosion-erosion. Water was flowing too fast.

Recommendation: Replace

Fig. 9

DISK BLADES

The following six disk blade illustrations represent the disk blade failures most often encountered.

These two disk blade failures resulted from defective material.

Recommendation: Replace

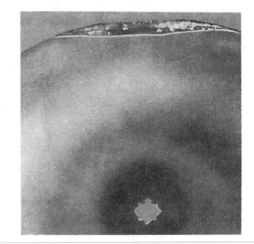

Fig. 10

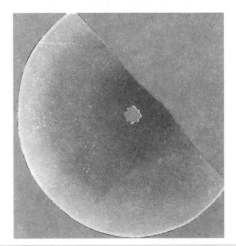

Fig. 11

Chipping caused by rock or stump conditions.

Recommendation: Replace

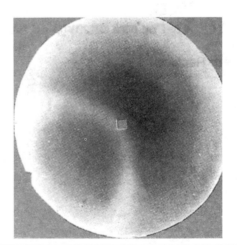

Fig. 12

These illustrate a non-directional break, resulting from rock or stump conditions.

Recommendation: Replace

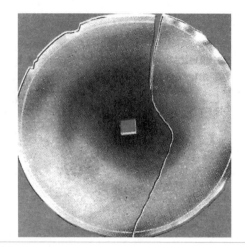

Fig. 13

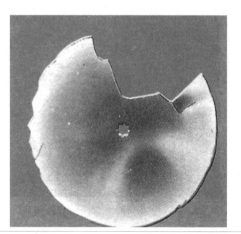

Fig. 14

This center was broken out by rock or stump conditions.

Recommendation: Replace

Fig. 15

PLANTER FINGER PICKUP

Replace the carrier plate when the chrome has worn off and the case-hardened steel begins to wear. Excessive wear on the dimples will cause over-population, especially with small corn seed.

Fig. 16

Lack or lubricant will accelerate wear of the carrier plate and the finger pickups. A teaspoon of powdered graphite should be sprinkled into each seed hopper daily.

Recommendation: Replace

Fig. 17

SPARK PLUGS

A spark plug with brown to grayish-tan deposits and slight electrode wear is normal and indicates good engine adjustments.

Fig. 18

Wet oil deposits with a minor degree of electrode wear may be caused by oil pumping past worn rings or excessive valve stem guide clearance.

"Break-in" of a new or recently overhauled engine before rings are fully seated may also result in this condition.

Recommendation: Clean, set electrode gap and reuse or replace

Fig. 19

Dry, fluffy and black carbon deposits result from over-rich carburetor adjustments. A clogged air cleaner can restrict air flow to the carburetor causing rich mixtures. Poor ignition output (faulty breaker points, weak coil or condenser) can reduce voltage and cause misfiring. A fouled spark plug is the result, not the cause of this problem.

Recommendation: Clean, reset electrode gap, and reuse or replace

Fig. 20

Red, brown, yellow, and white colored coatings which accumulate on the insulator are by-products of combustion and come from the fuel and lubricating oil, both of which today, generally contain additives.

Recommendation: Clean, reset electrode gap and reuse or replace

Fig. 21

Heat shock is a common cause of broken and cracked insulator tips. Incorrect ignition timing and a poor grade fuel are usually responsible for heat shock failures. Rapid increase in tip temperature under severe operating conditions causes the heat shock and fracture results.

Another common cause of chipped or broken insulator tips is carelessness in regapping by either bending the center wire to adjust the gap, or allowing the gapping tool to exert pressure against the tip of the center electrode or insulator when bending the side electrode to adjust the gap. Install a new spark plug.

Recommendation: Replace

Fig. 22

Pre-ignition causing burned or blistered insulator tip and badly eroded electrodes indicates excessive overheating. Clogged shrouding, dirty engine fins and sticky valves can also result in pre-ignition. Lean fuel-air mixtures are an additional cause.

Recommendation: Replace

Fig. 23

OIL TEST KITS

Oil test kits like the John Deere Oilscan kit can be used to check the condition of an engine, a separate transmission, or hydraulic system, gearbox, differential, or final drive. An oil sample from the area to be tested is analyzed for contamination and suspended particles.

Regular oil sampling of these areas can tell the operator of a machine that a problem may occur. The repairs can then be made before the potential problem can cause downtime.

The Oilscan kit is available through John Deere dealers. Other service locations and dealer organizations may offer similar services.

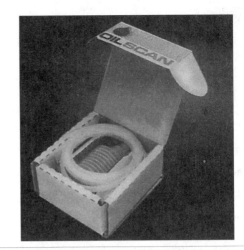

Fig. 24

O-RINGS

O-rings are used as seals in many components of a hydraulic system.

O-ring failure can be caused by:

- **Cuts or nicks from sharp objects**
- **Heat**
- **Improper fluid**
- **Lack of lubrication**
- **Improper installation**

Cuts or nicks occurring during installation of the O-ring or installation of the component is probably the most common cause of failure.

Excessive heat will damage O-rings as well as other seals. There are many reasons for the hydraulic system to overheat, but incorrect oil, low oil level, or dirty oil are the easiest to check and correct.

Using an incompatible fluid can not only cause a heat problem, it can create a chemical reaction which can soften or harden O-rings and other seals in the system.

It is recommended that O-rings always be replaced when reassembling a hydraulic component.

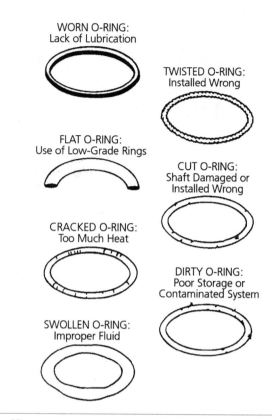

WORN O-RING:
Lack of Lubrication

TWISTED O-RING:
Installed Wrong

FLAT O-RING:
Use of Low-Grade Rings

CUT O-RING:
Shaft Damaged or
Installed Wrong

CRACKED O-RING:
Too Much Heat

DIRTY O-RING:
Poor Storage or
Contaminated System

SWOLLEN O-RING:
Improper Fluid

Fig. 25

OTHER HYDRAULIC SEALS

While the O-ring is probably the most used seal in the hydraulic system, there are a number of other seals that may be encountered. These include:

1. Cup packing

2. Flange packing

3. U-packing

4. V-packing

5. Spring-loaded lip seal

6. Compression packing

7. Mechanical seal

8. Non-expanding metallic seal

9. Expanding metallic seal

Most seals are fragile and can be easily damaged. Proper handling during storage and installation is critical. Don't damage the seal before it is put into use.

It is recommended that all seals be replaced whenever a component is repaired. Try to determine the cause of the leakage before disassembling the component. The problem may be something other than a damaged seal.

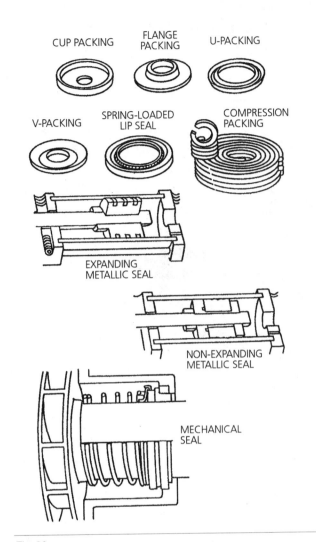

Fig. 26

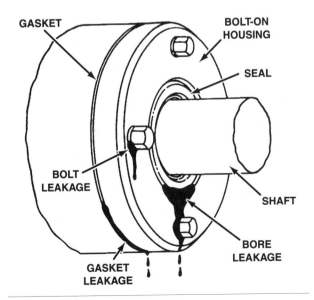

Fig. 27

If it is determined that the seal is damaged, try to find the cause.

Inspect the sealing surface or lips (bold arrow) of the seal for unusual wear, warping, cuts and gouges, or embedded foreign particles.

This seal has been worn by a rough shaft.

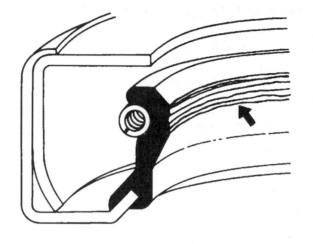

Fig. 28

Check shafts for roughness at seal contact areas. Deep scratches or nicks in the shaft can damage the seal.

A shaft spline, keyway, or burred end can also cause nick or cut in the seal lip during installation.

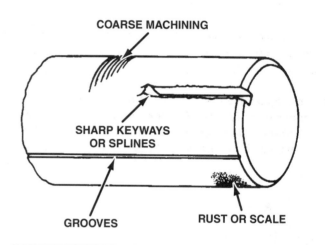

Fig. 29

Check the bore into which the seal is pressed. Imperfections in the bore can damage the seal causing leakage.

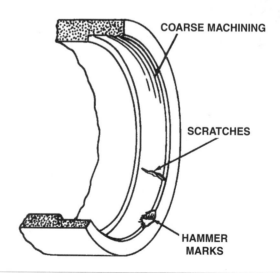

Fig. 30

The use of a non-recommended hydraulic fluid can either harden or soften the synthetic rubber of the seals.

If a factory approved seal has been used and the lips of the seal feels "spongy," an incompatible fluid has been used in the hydraulic system.

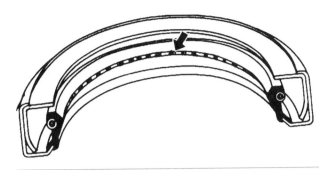

Hardening of the lips of the seal (bold arrow) could be from a chemical reaction caused by an incompatible fluid or excessive heat in the system.

Fig. 31

INTERCOOLER BRACE

Some intercoolers had braces (A) that could come loose and be pulled into the engine cylinder. The brace parts were embedded into the crown of the piston.

Recommendation: Replace with newer style intercooler.

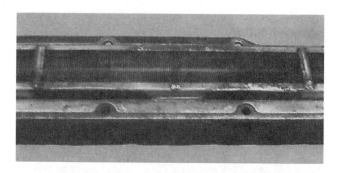

Fig. 32

FASTENERS

Most fastener failures occur while a machine is being used. Failures can also occur when a cap screw joint is tightened for the first time during assembly of a machine. Inadequate cap screw tension can result from improper assembly or unusual friction factors.

FAILURES DURING ASSEMBLY

BOLT TWIST OFF

This cap screw failure was from poor thread quality or a large friction factor between components.

Recommendation: Replace

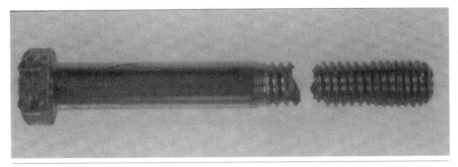

Fig. 33

GALLING

This deformed thread (galling) is due to poor thread contact with the nut.

Recommendation: Replace

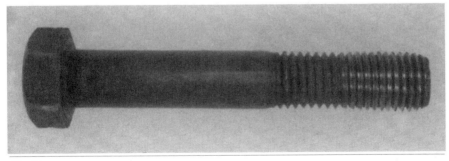

Fig. 34

BOLT YIELD AND TENSILE FAILURE

Excessive torque caused this cap screw to stretch and reduce the thickness of threads.

Recommendation: Replace and reduce assembly torque.

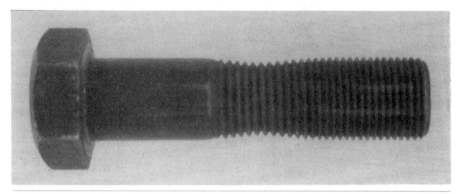

Fig. 35

When friction is low, the recommended torque will produce too much tension on the cap screw and cause this failure.

Recommendation: Replace and reduce torque.

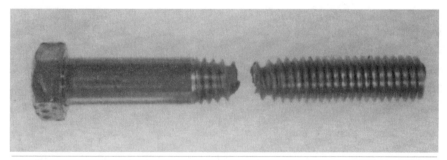

Fig. 36

NUT/CAP SCREW THREAD FAILURE

A cap screw can be "stripped" by using a nut with a short thread length or the incorrect hardness.

Recommendation: Replace with a higher grade nut, or increase nut height.

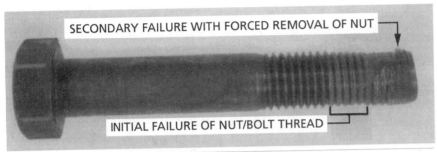

SECONDARY FAILURE WITH FORCED REMOVAL OF NUT

INITIAL FAILURE OF NUT/BOLT THREAD

Fig. 37

NUT DILATION

When a low grade nut is used in assembly, the washer face of the nut expands in diameter.

Recommendation: Replace with higher strength nut, add hard washer in joint or use a flanged nut.

Fig. 38

This shows fasteners with a higher strength than the components they are joining. The washer face of the nut is forced into the component surface. The hole in the component is deformed into the bolt.

Recommendation: Replace with a flanged nut, add hard washer or improve strength of components.

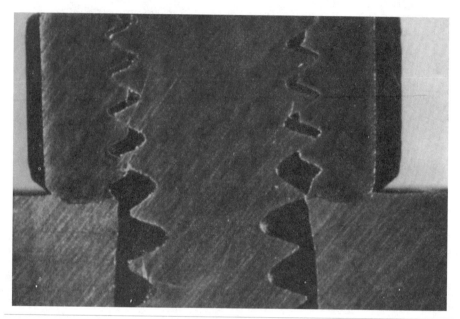

Fig. 39

GALLING

The galled surfaces of mating parts which rotate during assembly, are the result of poor surface finish or lack of lubricant.

Recommendation: Improve surface finish, add lubricant, or change hardness of surface.

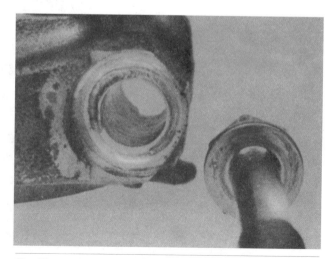

Fig. 40

FAILURES DURING USE OR AFTER ASSEMBLY

SHEAR FAILURE

This cap screw indicates a large crosswise load was placed on the joint components.

Recommendation: Replace and increase cap screw torque. Use bushings to carry shear loads. Replace with larger cap screw.

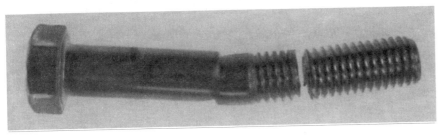

Fig. 41

FATIGUE FAILURE

These fasteners failed from a low cap screw torque, or a combination of low cap screw torque and high cyclis stress. The service loads may have been higher than expected, or compounded by bending loads.

Recommendation: Replace and use higher cap screw torque. Replace with larger cap screws. Reduce stress concentrations on cap screws.

FIRST THREAD ENGAGED IN NUT

Fig. 42

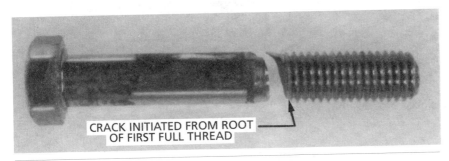

CRACK INITIATED FROM ROOT OF FIRST FULL THREAD

Fig. 43

LOSS OF CAP SCREW TORQUE AFTER ASSEMBLY

Dirty and rough surfaces can create vibration in the joint and cause self-loosening of the cap screw.

Recommendation: Clean surfaces to reduce contact stress. Use locking device in threads.

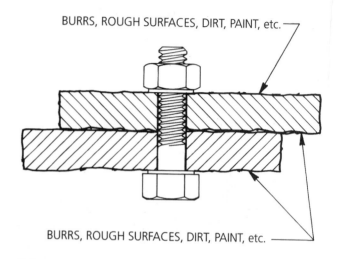

BURRS, ROUGH SURFACES, DIRT, PAINT, etc.

BURRS, ROUGH SURFACES, DIRT, PAINT, etc.

Fig. 44

TEST YOURSELF

QUESTIONS

1. (Fill in blanks.) A common cause of both pump rotor and impeller damage is _____ _____.

2. (Fill in blanks.) _____ _____ is another cause of impeller damage.

3. What is the most common cause of disk failures.

4. Black carbon deposits on the spark plugs indicate that the carburetor is adjusted too lean or too rich?

5. (Fill in blanks.) A common cause of broken or cracked insulator tips of spark plugs is _____ _____.

6. (True or false?) Heat shock failure of spark plugs is usually caused by incorrect ignition timing and a poor grade of fuel.

7. (Fill in blanks.) Oil pumping past worn piston rings or worn valve stem guides can cause _____ _____ _____ on sparkplugs.

8. (Fill in blank.) Overtightening cap screws or using too much grease during reassembly can cause O-rings to become _____.

9. (Fill in blanks.) _____ _____ can cause a cap screw to stretch and reduce the thickness of the threads.

10. (True or false?) Dirty and rough surfaces can cause vibration in the joint and cause a cap screw to loosen.

GLOSSARY OF TERMS

A

ABRASIVE WEAR
Surface damage caused by sliding contact with particles of hard foreign materials.

ADHESIVE WEAR
Surface damage commonly caused by metal-to-metal contact. Microscopic surface irregularities weld together, then tear apart resulting in scuffing or scoring of one or both of the surfaces in contact.

ANTI-FRICTION BEARINGS
Friction-reducing devices that use rolling contact components between mating surfaces such as balls or rollers.

B

BACKLASH (Gears)
The clearance, or "play" between two gears in mesh.

BANDED V-BELTS
Multiple V-belts permanently vulcanized to a tie band.

BEARINGS
See anti-friction bearings and journal bearings.

BEARING PLATES
A part in a hydrostatic transmission or hydraulic pump.

BLOW-BY
A leakage or loss of compression past the piston ring between the piston and the cylinder.

BORE
The internal surface of a cylindrical opening.

BRINELLING
The effect of one piece of metal being pressed into another, causing depressions to form on one or both contacting surfaces.

C

CAMSHAFT
The shaft containing lobes or cams to operate the engine valves.

CARBON DEPOSITS
Hard built-up deposits on surfaces formed on engine parts exposed to gaseous products of combustion.

CARBURIZATION
The addition of carbon to the surface of steel parts, by heat treatment, to provide high hardness for improved wear resistance and durability. A form of case-hardening often applied to highly loaded gears.

CASE CRUSHING
Crushing of the outer surface (case) of a gear that has been case-hardened by heat treatment.

CAVITATION DAMAGE
Pitting of a metal surface exposed to a moving liquid by the collapse of entrained vapor bubbles. Corrosion is often a factor.

CHAINS
A flexible series of metal links, or rings, fitted into one another.

CHEMICAL CORROSION
Surface damage caused by contact with a chemically active liquid or gas — such as rusting of steel in contact with moist air or water.

COLD FLOW
Metal movement, under high pressure, at room temperature.

COMBUSTION KNOCK
A noise that occurs when fuel in a cylinder is ignited too early, too rapidly, or unevenly.

CONTAMINATION
Foreign material that could damage a part.

CORROSION
See chemical corrosion.

CRANKSHAFT
The main drive shaft of an engine which converts reciprocating motion to rotary motion by means of cranks.

CRUSH
A condition in journal bearings. Each half of the bearing insert must extend a very small amount beyond the parting edges of the rod and rod cap. When the capscrews or rod bolts are tightened, the bearing halves are pressed into the bore for a snug fit.

CYLINDER BLOCK
The housing of an engine, a hydraulic pump or a hydraulic motor containing the cylinder bores along with other functional components.

CYLINDER BORE
The inside surface of the passage in a cylinder block in which a piston moves.

CYLINDER BORE BUSHING
A sleeve or tube between the piston and cylinder block in a hydrostatic transmission.

CYLINDER HEAD
That part of an engine bolted to the cylinder block, forming the closed end of the combustion chamber. It contains the valves and the passages for fuel, air, exhaust gases and cooling water.

CYLINDER LINER
A replaceable sleeve or tube inserted in the cylinder block to provide a renewable cylinder bore in engines.

D

DETONATION

Uncontrolled combustion accompanied by a loss of power and waste of energy.

DRIVE BELT

A belt used to transmit power between a driving and a driven pulley.

E

ELECTRICAL PITS

Removal of a small portion of surfaces in contact by an electrical current.

EROSION

The wearing away of a surface by the impingement of abrasive particles suspended in a gas or liquid.

F

FATIGUE

Failure of a part initiated by cracking resulting from repeated subjection to stresses too low to cause failure in a single application.

FIT RUST

Rusting that results when the outer race of an anti-friction bearing fits too loose in its housing. A particular form of fretting (See Fretting).

FLAKING

The separation of thin layers from the surface of a metal part.

FOREIGN MATERIAL

Any substance that is in a place where it is not supposed to be. An example is dirt in lubricating oil.

FRETTING

A type of wear resulting from slight reciprocating movement between metal parts that are in close contact. Generally accompanied by corrosion or rusting.

G

GALLING

Surface damage on mating, moving metal parts due to friction. A severe form of adhesive wear. (See Adhesive Wear).

GEAR

A cylinder or cone-shaped part having teeth on one surface which mate with and engage the teeth of another part which is not on the same axis.

GLOW PLUG

A device designed to improve starting conditions in some diesel engines. A heating element in the combustion chamber or intake manifold.

GOUGING

Severe grooving of a surface by sliding or impact of large, hard particles.

GROOVE

A long, narrow channel in a surface.

H

HYDROSTATIC TRANSMISSION

A hydraulic transmission which uses fluid under pressure to transmit engine power to the drive wheels of the machine.

I

IMPACT FAILURE

The failure of a part that results from a single blow or suddenly applied force sufficient to cause an immediate fracture.

J

JOURNAL BEARING

Bearings that provide a sliding contact between mating surfaces; bushings.

K

KNOCKING

A noise that occurs when fuel in a cylinder is ignited too early, too rapidly, or unevenly.

L

LINK (Track and Undercarriage)

A section of a track. A series of links, connected in chain-like manner by bushings and pins, attached to the respective track shoes make up the track.

LUBRICATION

Use of a substance (grease, oil, etc.) to reduce friction between parts or objects that move against each other.

LUGGING

Occurs when an engine operating at a specified rpm is given a load it cannot support at the same, or a higher, rpm.

O

OIL RING

The piston ring nearest the bottom of the piston that controls lubrication between the piston and cylinder liner.

OVERFUELING

Allowing more fuel to enter the engine combustion chamber (thus greater stress on parts) than the engine was designed to handle.

OVERHEATING

Allowing the temperature of an object to heat to a higher temperature than it was designed to handle.

OVERLOADING

Allowing the weight that an object carries or pulls to become greater than the object was designed to handle.

OVERSPEEDING

Allowing the rpm of an engine to accelerate to a higher level than the engine was designed to handle.

P

PINION

The smaller of two meshing gears.

PISTON

A cylindrical part closed at one end which is connected to the crankshaft by the connecting rod. The force of the expanding gases in the cylinder is exerted against the closed end of the piston causing the connecting rod to move the crankshaft.

PISTON RING

An expanding ring placed in the grooves of the piston to seal off the passage of fluid or gas past the piston.

PITTING (Gears or Bearings)

A type of surface damage occurring under repeated loading of two parts in rolling or sliding contact. A form of surface fatigue.

PLANET GEAR

Gears in a planetary gear set which connect the sun gear to the ring gear.

PREIGNITION

Ignition occurring earlier than intended (gasoline engine). For example, the fuel-air mixture being ignited in a cylinder by an extremely hot particle in the combustion chamber before the electrical spark occurs.

PUMP IMPELLER

The rotating member in a pump that provides continuous power to move fluids.

PUMP ROTOR

Similar to pump impeller.

PUSH ROD

A tubular or solid rod for transmitting motion between the valve lifters and rocker arms in an internal combustion engine. The push rod is actuated by the lobes on the camshaft to open and close valves.

R

RIDGING

Deformation of the working surface of gear teeth caused by high stress between two gears in mesh.

RING GEAR

A ring shaped part having teeth on either the outside or inside diameter. Examples of this are the starter ring gear, the gear which surrounds the sun and planet gears in a planetary system or the spiral bevel gear in a differential.

RIPPLING

A periodic wave-like formation on gear teeth caused by high stress between two gears in mesh.

ROCKER ARM

In an internal combustion engine, a lever that is pivoted near its center operated by a push rod at one end to depress the valve stem at the other end.

ROLLERS (Track and Undercarriage)

Rolling members of a track that prevents the track from sagging on the top by supporting the weight of the upper portion of the track. Rollers mounted on the bottom of the frame support much of the machine's weight.

ROTATING ASSEMBLY (Turbocharger)

The parts in a turbocharger that turn or rotate (wheels and shaft).

RUPTURE (Belt)

The surface appearance of a broken belt.

S

SCORING

Grooving of the surface of one or both parts in relative motion to each other by particles of a hard foreign material which may be free or embedded in one of the surfaces.

SCUFFING

A form of surface damage caused by intermittent lack of lubrication (see Adhesive Wear).

SEIZURE

The sudden stopping of motion between two parts because of excessive friction and heat between parts. Heat causes the parts to expand, reducing clearance between the two parts to zero.

SHAFT

A long, slender, usually cylindrical part.

SHOCK LOADING

Placing an extremely heavy load on a part for a short period of time.

SLUDGE

A precipitate from oils, such as products from crankcase oils in engines.

SMEARING

The displacement of surface metal from one area of the surface to another; usually caused by another part rubbing on the surface with insufficient clearance and lubrication between parts.

SPALLING

Breaking off of small particles from the surface of a part. Repeated stress from heavy contact between two metal surfaces (such as gears in mesh) causes spalling. An advanced stage of pitting.

SPARK PLUG

A device that screws into the cylinder of an internal combustion engine (gasoline) having a pair of electrodes between which an electrical discharge is passed to ignite the fuel/air mixture.

SPINDLE

A short, slender, tapered shaft.

SPROCKET

A wheel having teeth which engage the links of a chain for the purpose of transmitting power.

STATIC FAILURE

Failure of a part because the load was too high (overloading).

STEERING KNUCKLE

The hinged or articulated assembly for transmitting force from the steering spindle to the front axle for controlling direction of travel.

STRESS

The load per unit area.

SWASHPLATE

A plate in a hydrostatic transmission that controls the position and movement of the pistons and thus determines the speed and direction in which a machine moves.

T

TAPPET

The adjusting device for varying the clearance between the valve stem and the cam. May be built into the valve lifter in an engine or may be installed in the rocker arm on an overhead valve engine.

TORQUE

The effort of twisting or turning.

TORSIONAL FATIGUE

The cracking or failure of a part due to continuous or repeated twisting on the part at a force too great for the part to withstand.

TRACK (Crawler)

An assembly consisting of ground engaging shoes attached to track links, driven by a sprocket, to propel a crawler vehicle.

TRANSMISSION

An assembly of gears or other elements used to obtain variations in speed or direction between the input and output shafts.

TURBOCHARGER

An exhaust-driven turbine which drives a centrifugal compressor wheel.

U

UNIVERSAL JOINT

A linkage that transmits rotation from one shaft to another not collinear with it.

V

VALVE

Any movable device for controlling the movement of a liquid or gas by opening or closing a passage. In an engine, the device for opening and closing the cylinder intake and exhaust ports.

VALVE GEAR TRAIN

The engine valves and valve operating mechanism that controls fuel and air flow to and from the combustion chamber.

VALVE SEAT

The matched surface upon which the valve face rests.

VALVE SPRING

A spring attached to a valve to return it to its seat after it has released from the lifting or opening means.

VALVE STEM GUIDE

A bushing or hole in which the valve stem slides.

V-BELT

Belts that transmit power between drive and driven pulleys or sheaves. The area of the belt that fits in the sheave groove is shaped like a "V" so it can be wedged against the sides of the sheave groove to provide enough friction to transmit power.

VIBRATION

A quivering or trembling motion.

ANSWERS TO ' TEST YOURSELF ' QUESTIONS

CHAPTER 1

1. a) Lean fuel mixture
   b) fuel octane too low
   c) ignition timing advanced too much
   d) lugging the engine or overfueling
   e) cooling system overheating

2. Preignition

3. Scuffing and scoring

4. Abrasives entering the system from external sources.

5. Deposits caused by too much heat, unburned fuel, and excess lubricating oil first collecting and then hardening causing the rings to stick in their grooves.

6. a) Forms gas-tight seal between the piston and cylinder
   b) helps to cool the piston by transferring heat
   c) controls lubrication between the piston and cylinder wall

7. c) Cylinder liner

ANSWERS TO ' TEST YOURSELF ' QUESTIONS

CHAPTER 2

1. Dirt

2. a) Clean area surrounding the bearing thoroughly during installation
   b) follow recommended maintenance for air and oil filters

3. Applying the recommended torque to the rod bolts.

4. A misaligned connecting rod.

5. Corrosion from acid formation in the oil.

ANSWERS TO ' TEST YOURSELF ' QUESTIONS

CHAPTER 3

1. The valves, particularly the exhaust valves.

2. Too much heat.

3. Excessive heat in the valve stem; heat not dissipating from the stem through the valve guide to the cylinder block.

4. Incorrect tappet clearance causing valve to be held off its seat.

5. True.

6. True.

7. Heat fatigue from overheating.

8. A buildup of carbon particles between the valve and its seat.

9. The seating of the valve with too much force, often caused by too much valve clearance.

10. False

ANSWERS TO ' TEST YOURSELF ' QUESTIONS

CHAPTER 4

1. a) Foreign material
   b) contact damage
   c) erosion or abrasion

2. To prevent oil starvation of turbocharger shaft and bearings

3. Contaminated oil

4. Extreme temperatures

5. False

6. Lack of or contaminated lubricant and foreign material in the assembly

ANSWERS TO ' TEST YOURSELF ' QUESTIONS

CHAPTER 5

1. Adhesive, abrasive, and corrosive

2. Excessive operating loads

3. Fatigue

4. Impact

5. Overloading of the gear

6. True

7. Improper lubricant, low lubricant level, or infrequent lubricant change

8. False

ANSWERS TO ' TEST YOURSELF ' QUESTIONS

CHAPTER 6

1. The weight the shaft must support while sitting still

2. True

3. Overloading

4. True

5. Lack of or contaminated lubricant

6. Breakdown of lubricating film

ANSWERS TO ' TEST YOURSELF ' QUESTIONS

CHAPTER 7

1. Fine scratches or grooves on a metal surface

2. Abrasive contaminants in the hydraulic fluid

3. When the scratches or grooves can be felt by the fingernail or lead pencil

4. A metal transfer created by severe friction between rotating parts

5. Lack of lubrication

6. Pitting or wearing away of a metal surface

7. Air bubbles in the fluid collapsing or exploding against the surface

8. Improper fluid, fluid contaminated with water, excessive temperature

9. Abrasive contaminants in the hydraulic fluid

10. Lack of lubrication and contaminants in the fluid.

ANSWERS TO ' TEST YOURSELF ' QUESTIONS

CHAPTER 8

1. Moisture and abrasives like dirt and sand

2. Adequate use of the correct lubricant

3. False

4. True

5. False (creep wear)

6. True

7. A bent shaft or foreign material between the bearing and its seat

8. True

ANSWERS TO ' TEST YOURSELF ' QUESTIONS

CHAPTER 9

1. True

2. a) Badly worn sheaves
   b) Too much tension forcing the belt down into the sheave grooves
   c) An object failing onto the sheave groove while the drive is operating

3. The belt comes in contact with some part of the machine; most likely because the belt is running too loose

4. Slipping under a starting or stalling load

5. The belt climbing out of the sheave groove during operation

6. Foreign object in the sheave groove

7. Lack of lubrication

8. False

ANSWERS TO ' TEST YOURSELF ' QUESTIONS

CHAPTER 10

1. Sprockets becoming packed with dirt or clay changing the sprocket and tooth pitch

2. True

3. False

4. True

5. Hitting an object

6. Over-inflation of the fires

7. True

8. False (some can be repaired)

ANSWERS TO ' TEST YOURSELF ' QUESTIONS

CHAPTER 11

1. Abrasive wear

2. Cavitation corrosion

3. Rock or stump conditions in the field

4. Too rich

5. Heat shock

6. True

7. Wet oil deposits

8. Extruded

9. Excessive torque

10. True

INDEX